Stellar Evolution
and its Relations
to Geological Time

JAMES CROLL

CAMBRIDGE
UNIVERSITY PRESS

CAMBRIDGE UNIVERSITY PRESS

Cambridge, New York, Melbourne, Madrid, Cape Town,
Singapore, São Paolo, Delhi, Mexico City

Published in the United States of America by Cambridge University Press, New York

www.cambridge.org
Information on this title: www.cambridge.org/9781108048361

© in this compilation Cambridge University Press 2012

This edition first published 1889
This digitally printed version 2012

ISBN 978-1-108-04836-1 Paperback

STELLAR EVOLUTION

PRINTED BY EDWARD STANFORD

26 & 27 COCKSPUR STREET, CHARING CROSS, LONDON, S.W.

STELLAR EVOLUTION

AND ITS

AND ITS

RELATIONS TO GEOLOGICAL TIME

BY

JAMES CROLL, LL.D., F.R.S.

AUTHOR OF 'CLIMATE AND TIME' 'CLIMATE AND COSMOLOGY'
'PHILOSOPHY OF THEISM' ETC.

LONDON : EDWARD STANFORD

26 & 27 COCKSPUR STREET, CHARING CROSS, S.W.

1889

PREFACE.

THERE are two, and only two, conceivable sources from which the prodigious amount of energy possessed by our sun and solar system can possibly have been derived. Not only are these two sources radically distinct in their essential nature, but both are admitted to be real and not merely hypothetical sources of energy. The one source is gravitation; the other, the source discussed in the present volume, a source to which attention was directed some twenty years ago. A most important distinction between these two sources is this: the amount of energy available from the former can be accurately determined, but such is not the case in regard to the latter. We can tell with tolerable certainty the greatest amount of energy which gravitation could possibly have conferred on the sun and solar system; but we have, at present, no means of assigning a limit to the possible amount which might have been derived from the other source. It may have been equal to that which gravitation

could afford, or it may have been twofold, fourfold, or
even tenfold that amount.

We have evidently in this case a means of determining which of the two sources will ultimately have
to be adopted as the source to which the energy of our
solar system must be referred. For if it can be proved
from the admitted facts of geology, biology, and other
sciences, that the amount of energy in the form of
heat which has been radiated into space by the sun
during geological time is far greater than the amount
which could possibly have been derived from gravitation, this will undoubtedly show that gravitation cannot account for the energy originally possessed by our
system.

The First Part of the volume is devoted to the
consideration of what I believe to be the probable
origin of meteorites, comets, and nebulæ, and of the
real source from which our sun derived his energy.
The facts which support the theory here advocated,
together with the light which that theory appears to
cast upon those facts, are next considered ; and it will
be found, I think, that the theory has been very much
strengthened by the recent important spectroscopic
researches of Mr. Lockyer and others in reference to
the constitution of nebulæ. The Second Part of the
work deals with the evidence in support of the theory
derived from the testimony of geology and biology as
to the age of the sun's heat. The Third, and last,

Part has been devoted to questions relating to the pre-nebular condition of the universe, and the bearing which these have on theories of stellar evolution. Several subjects introduced in this part are only very briefly treated. These will, however, be considered at greater length in a future volume, "Determinism, not Force, the Foundation-stone of Evolution," a work of a more general and abstract character, which was commenced many years ago.

PERTH: *January* 2, 1889.

CONTENTS.

—◦◦—

PART I.

THE IMPACT THEORY OF STELLAR EVOLUTION.

a

PART II.

EVIDENCE IN SUPPORT OF THE THEORY FROM THE AGE OF THE SUN'S HEAT.

PART III.

EVIDENCE IN SUPPORT OF THE THEORY FROM THE PRE-NEBULAR CONDITION OF THE UNIVERSE.

CONTENTS

STELLAR EVOLUTION.

PART I.

THE IMPACT THEORY OF STELLAR EVOLUTION.

UPWARDS of twenty years ago [1] the theory—or, I should rather say, the hypothesis—was advanced [2] that our sun was formed from a hot gaseous nebula produced by the colliding of two dark stellar masses ; and that, as the stars are suns like our own, they in all likelihood had a similar origin. The probability of this theory has been very much strengthened by the facts, both astronomical and physical, which have accumulated since the theory was enunciated. Before proceeding to the consideration of these facts, and the conclusions to which they lead, it will be necessary to give a statement of the fundamental principles of the theory.

In the theory here discussed the truth of the nebular hypothesis, which begins by assuming the

---

[1] *Philosophical Magazine*, May 1868 ; *Climate and Time*, chap. xxi.; *Quarterly Journal of Science*, July 1877 ; *Phil. Mag* July 1878 ; *Climate and Cosmology*, chaps. xvii. xviii. and xix.

[2] I prefer to use the term " theory," with the above understood qualification, viz. a theory in its hypothetical stage.

existence of a solar nebulous mass, is taken for granted. The present theory deals not so much with the nebulous mass itself as with the *formation* of the nebula, and with those causes which led to its formation. For convenience of reference, and to prevent confusion, I have called it the "Impact Theory," by which name it may be distinguished, on the one hand, from the nebular theory, and, on the other hand, from the meteoric theory, and all other theories which regard gravitation as the primary source of the solar energy.

The theory starts with the assumption that the greater part of the energy possessed by the universe exists or is stored up in the form of the motion of stellar masses. The amount of energy which may thus be stored up is startling to contemplate. Thus a mass equal to that of the sun, moving with a velocity of 476 miles per second, would possess, in virtue of that motion, energy sufficient, if converted into heat, to maintain the present rate of the sun's radiation for 50,000,000 years.[1] There is nothing extravagant in the assumption of such a velocity. A comet, for example, having an orbit extending to the path of the planet Neptune, approaching so near the sun as to almost graze his surface in passing, would have a velocity within 86 miles of what we have assumed. Twice this assumed velocity would give 200,000,000 years' heat; four times the velocity would give 800,000,000 years' heat; and so on.

We are at perfect liberty to begin by assuming the

---

[1] Pouillet's estimate of the rate of solar radiation is here taken.

existence of stellar masses in motion ; for we are not called upon to explain how the masses obtained their motion, any more than we have to explain how they came to have their existence. If the masses were created, they may as likely have been created in motion as at rest ; and if they were eternal, they may as likely have been eternally in motion as eternally at rest.

Eternal motion is just as warrantable an assumption as eternal matter. When we reflect that space is infinite—at least in thought—and that, for aught we know to the contrary, bodies may be found moving throughout its every region, we see that the amount of energy may be perfectly illimitable.

But, illimitable as the amount of the energy may be, it could be of no direct service while it existed simply as the motion of stellar masses. The motion, to be available, must be transformed into heat : the motion of translation into molecular, or some other form of motion. This can be done in no other way than by arresting the motion of the masses. But how is such motion to be arrested ? How are bodies as large as our earth, moving at the rate of hundreds of miles per second, to have their motion stopped ? According to the theory this is effected by *collision* : by employing the motion of the one body to arrest that of the other.

Take the case of the formation of our sun according to the theory. Suppose two bodies, each one-half of the mass of the sun, moving directly towards each other with a velocity of 476 miles per second. These bodies would, in virtue of that velocity, possess 4149×10^{38} foot-pounds of energy, which is equal to 100,000,000,000 foot-pounds per pound of the mass ; and this, converted into heat by the stoppage of their motions, would suffice to maintain, as was previously stated, the present rate of the sun's radiation for a period of 50,000,000 years. It must be borne in mind that, while 476 miles per second is the velocity at the moment of collision, more than one-half of this would be derived from the mutual attraction of the two bodies in their approach to each other.

Coming in collision with such a velocity, the result would inevitably be that the two bodies would shatter each other to pieces. But, although their onward motions would thus be stopped, it is absolutely impossible that the whole of the energy of their motions could be at once converted into heat ; and it is equally impossible that it could be annihilated. Physical considerations enable us to trace, though in a rough and general way, the results which would necessarily follow. The broken fragments, now forming one confused mass, would rebound against one another, breaking up into smaller fragments, and flying off in all directions. As these fragments receded from the centre of dispersion they would strike against each other, and, by their mutual impact, become shivered into still smaller fragments, which would in turn be broken up into fragments yet smaller, and so on as they proceeded outwards. This is, however, only one part of the process, and a part which would certainly take place, though no heat were generated by the collisions.

A far more effective means of dispersing the fragments and shattering them to pieces would be the expansive force of the enormous amount of incandescent gas almost instantaneously generated by the heat of collision. The general breaking up of the two masses and the stoppage of their motions would be the work of only a few minutes, or a few hours at most. The heat evolved by the arrested motion would, in the first instance, be mainly concentrated on the surface layers of the broken blocks. The layers would be at once transformed into the gaseous condition, thus enveloping the blocks and filling the interspaces. It is difficult to determine what the temperature and expansive force of this gas would at the moment be, but evidently it would be excessive ; for, were the whole of the heat of the arrested motion distributed over the mass, it would, as has been stated, amount to 100,000,000,000 foot-pounds per pound of the mass— an amount sufficient to raise 264,000 tons of iron 1° C. Thus, if we assume the specific heat of the gas to be equal to that of air (viz. 2374), it would have a temperature of about 300,000,000° C. or more than 140,000 times that of the voltaic arc.

I hardly think it will be deemed extravagant to assume that at the moment after impact the temperature of the evolved gas would be at least as great as here stated. If we assume it to be so, it is obvious that the broken mass would, by the expansive force of the generated gas, be dispersed in all directions, breaking up into fragments smaller and smaller as they knocked against one another in their progress outwards from the centre of dispersion; and these fragments would, at the same time, become gradually converted into the gaseous state, and gradually come to occupy a space as large as that embraced in our solar system. In the course of time the whole would assume the gaseous condition, and we should then have a perfect nebula—intensely hot, but not very luminous. As its temperature diminished, the nebulous mass would begin to condense, and ultimately, according to the well-known nebular hypothesis, pass through all the different phases of rings, planets, and satellites into our solar system as it now exists.

I am glad to find that the theory, in one of its main features, has been adopted by Sir William Thomson,[1] the highest authority we have on all points relating to the source of the sun's heat.

"We cannot," says Sir William, "help asking the question, What was the condition of the sun's matter before it came together and became hot? (1) It may have been two cool, solid masses, which collided with the velocity due to their mutual gravitation; or (2), but with enormously less of probability, it may have been two masses colliding with velocities considerably greater than the velocities due to their mutual gravitation."

[1] Lecture on "The Probable Origin, the Total Amount, and the Possible Duration of the Sun's Heat," delivered at the Royal Institution on January 21, 1887, and published in *Nature* of 27th of the same month. The lecture was afterwards published with considerable additions and alterations in the *Proceedings of the Institution*, vol. xii. It is from this that my quotations are taken.

He adopts the first of these suppositions. "To fix the idea," he continues, "think of two cool, solid globes, each of the same mean density as the earth, and of half the sun's diameter, given at rest, or nearly at rest, at a distance asunder equal to twice the earth's distance from the sun. They will fall together and collide in exactly half a year. The collision will last for about half an hour, in the course of which they will be transformed into a violently agitated incandescent fluid mass flying outward from the line of the motion before the collision, and swelling to a bulk several times greater than the sum of the original bulks of the two globes. How far the fluid mass will fly out all around from the line of collision it is impossible to say. The motion is too complicated to be fully investigated by any known mathematical method; but with sufficient patience a mathematician might be able to calculate it with some fair approximation to the truth. The distance reached by the extreme circular fringe of the fluid mass would probably be much less than the distance fallen by each globe before the collision, because the translational motion of the molecules constituting the heat into which the whole energy of the original fall of the globes becomes transformed in the first collision is probably about three-fifths of the whole amount of that energy. The time of flying out would probably be less than half a year, when the fluid mass must begin to fall in again towards the axis. In something less than a year after the first collision the fluid will again be in a state of maximum crowding round the centre, and this time probably

even more violently agitated than it was immediately after the first collision ; and it will again fly
outward, but this time axially towards the places
whence the two globes fell. It will again fall in-
wards, and after a rapidly subsiding series of quicker
and quicker oscillations it will subside, probably in
the course of two or three years, into a globular star
of about the same dimensions, heat, and brightness
as our present sun, but differing from him in this,
that it will have no rotation." [1]

This is precisely what I have been contending
for during the past twenty years, with the simple
exception that I assume, according to his second supposition, that the " two masses collided with velocities
considerably greater than the velocities due to mutual
gravitation." Sir William admits, of course, my supposition to be quite a possible one, but rejects it on
the supposed ground of its improbability. His reasons
for this, stated in his own words, are as follows :

" This last supposition implies that, calling the
two bodies A and B for brevity, the motion of the
centre of inertia of B relatively to A must, when the
distance between them was great, have been directed
with great exactness to pass through the centre of
inertia of A ; such great exactness that the rotational
momentum or moment of momentum after collision
was no more than to let the sun have his present
slow rotation when shrunk to his present dimensions.
This exceedingly exact aiming of the one body at the
other, so to speak, is, on the dry theory of probability,

exceedingly improbable. On the other hand, there is certainty that the two bodies A and B at rest in space if left to themselves, undisturbed by other bodies and only influenced by their mutual gravitation, shall collide with direct impact, and therefore with no motion of their centre of inertia, and no rotational momentum of the compound body after the collision. Thus we see that the dry probability of collision between two neighbours of a vast number of mutually attracting bodies widely scattered through space is much greater if the bodies be all given at rest than if they be given moving in any random directions and with any velocities considerable in comparison with the velocities which they would acquire in falling from rest into collision."

Sir William here argues that the second supposition is far less probable than the first, because, according to it, the motion of the one body relatively to the other must, in order to strike, be directed with great exactness. The result, in such a case, is that collision will rarely occur; whereas, according to the first supposition, the two bodies starting from a state of rest will, by their mutual gravitation, inevitably collide. According to the second hypothesis they will generally miss; according to the first they will always collide.

I have been led to a conclusion directly opposed to Sir William's. The fact that, according to the second supposition, collisions can but rarely occur is one reason, amongst others, why I think that supposition to be true ; and the fact that, according to the first

supposition, collisions must frequently occur is also
one reason, amongst others, why I think it very improbable that it can represent the true condition of
things.

It by no means adds anything to the probability
of the first supposition to assert that, according to it,
such collisions will occur readily and frequently.   On
the contrary, it would show that the supposition was
the less likely to be true.   If the collisions were insufficient in character, the fewer of them that occurred,
the better ; for the result of such collisions would
simply be a waste of the potential energies of the universe.   We should in this case have an innumerable
host of imperfect suns without planets, or with at
most only one or two, and these at no great distance
from the luminary.   There would thus be evolved
a universe without any grand planetary systems.
There is still another objection to the supposition.
The same gravitating force which makes the dark
bodies liable to come into collision with each other
must, of course, make them equally liable to come
into collision with the luminous bodies, and with our
sun amongst the rest.   Our sun would, accordingly,
be at the mercy of any of those masses which might
happen to come within the reach of its attractive influence.   It would pull the mass towards it, and a
collision would be inevitable, unless it so happened
that a transverse motion of the sun itself might enable
it to escape destruction.   Even in such a case it could
not by any means manage to get rid of the entangling
mass.

All this risk, in so far as gravitation is concerned, would have been completely averted if an original projected velocity of some thirty or forty miles per second had been conferred on the dark mass; for, in this case, the attractive force of the sun would fail to arrest its motion, and the mass would pass onward through space, never to return. This simple conception of an original motion removes entirely those objections which, we have seen, besets the supposition we have been considering. With such a motion, not only would the risk to our solar system be removed, but the collisions between the dark bodies themselves would be a matter of rare occurrence; and hence the energy of the universe would be conserved. And when a collision did happen it would be on a grand scale, and the result would be not an imperfect sun without planets, but an incandescent nebula, out of which, by condensation, a complete solar system would be evolved. In fact, within the whole range of cosmical physics, I know of nothing more impressive in its sublime simplicity than this plan, by which the stability and perfection of the universe is thus secured. How vast the ends—how simple the means!

Consideration of the Facts which support the Theory, and of the Light which the Theory appears to cast upon the Facts.

I. *Probable Origin of Meteorites.*

Recent researches establish beyond doubt that stars, nebulæ, comets and meteorites, do not differ much from our earth in their chemical constitution. Meteorites, it is true, differ in their physical characteristics from ordinary rock such as is found on the earth's surface. But it is possible, if not probable, that the earth's interior mass " may," as Sir Henry Roscoe remarks, " partake of the physical nature of these metallic meteorites, and that if we could obtain a portion of matter from a great depth below the earth's surface we should find it exactly corresponding in structure as well as in chemical composition with a metallic meteorite, and the existence of such interior masses of metallic iron may go far to explain the well-known magnetic condition of our planet." [1]   I think there can be little doubt that, were our earth broken up into small fragments, and these scattered into space, it would probably be impossible to distinguish them from ordinary meteorites.   The two would be so like in character that one can hardly resist the conviction that meteorites are but the fragments of sidereal masses which have been shattered by collision.   That meteorites are broken fragments is the opinion expressed by Sir William Thomson, who

[1] *Manchester Science Lectures*, Fifth Series, p. 31.

says "that he cannot but agree with the common opinion which regards meteorites as fragments broken from larger masses, and that we cannot be satisfied without trying to imagine what were the antecedents of those masses." The theory we have been considering appears to afford an explanation of their antecedents. According to it, they are broken fragments of two dark stellar masses which were shattered to pieces by collision. After what has been stated concerning the production of the gaseous nebulæ out of which our solar system was formed, it must be regarded as highly improbable, if not impossible, that the whole of the fragments projected outwards with such velocity should be converted into the gaseous condition. Multitudes of the smaller fragments, especially those towards the outer circumference of the nebulous mass, meeting with little or no obstruction to their onward progress, would pass outwards into space with a velocity which would carry them beyond the risk of falling back into the nebula. They would then continue their progress in their separated forms as meteorites. If this be their origin, then meteorites are the offspring of sidereal masses, and not their parents, as Mr. Lockyer concludes.

These meteorites must be of vast antiquity, for if they are fragments of the dark bodies then they must be not only older than our solar system, but older than the nebula from which that system was formed. Some of them, however, may have come from other systems. They are fragments which may yet cast some light on the history of the dark bodies.

Comets, bodies which in many points seem allied to meteorites, probably have, as we shall shortly see, a similar origin.

II. *Motion of the Stars ; how of such different velocities, and always in straight lines.*

It will be only when the two bodies, coming from contrary directions, collide with equal momentum that the entire motion will be stopped. But in the case of stellar masses moving, as it were, at random in every direction this is a condition which will but rarely occur. Accordingly, in most cases the resulting stars will have more or less motion. In short, the stars should, according to the theory, be moving in all directions and with all varieties of velocity. Further, it follows that these motions ought to be in perfectly straight lines, and not in definite orbits of any kind. So far as observation has yet determined, all these conditions seem to be fulfilled. Sometimes it will happen that the two bodies will strike each other obliquely. In this case the resulting star, both as to the direction and velocity of its motion, will, to a large extent, be the resultant of the two concurrent forces.

III. *Motion of the Stars not due to their Mutual Attractions.*

According to the theory the absolute motion of the stars is due, not to the influence of gravity, but to motions which originally belonged to the two component masses out of which the star arose ; motion

regarding the origin of which science can no more in-
form us than it can regarding the origin of the masses
themselves. There is strong presumptive evidence
that the motion of the stars is due to this cause. We
know that there are stars which have a far greater
velocity than can result from gravitation, such, for
example, as the star 1830 Groombridge, which has
a velocity of 200 miles per second. Suppose, with
Professor Newcomb, that the number of stars belonging to the universe amounts to 100,000,000, and that
these have, on the average, five times the mass of the
sun, and are spread out in a layer across which light
requires 30,000 years to pass. Then computation
shows that, unless the attractive power of the whole
were sixty-four times greater than it really is, it could
not have conferred on Groombridge the motion which
it possesses, or arrest it in its onward course.[1] We
are therefore forced, as Professor Newcomb remarks,
to one of two alternatives, viz.: "Either the bodies
which compose our universe are vastly more massive
and numerous than telescopic examination seems to
indicate, or 1830 Groombridge is a runaway star,
flying on a boundless course through infinite space,
with such momentum that the attraction of all the
bodies of the universe can never stop it."

As regards the theory we are discussing, it is the
same which alternative is taken, for both are equally
favourable. If the former, then, according to the
theory that stellar heat had its origin in collision, it is
presumptive evidence that space is occupied by dark

[1] Newcomb's *Astronomy*, p. 487, English edition, 1878.

bodies far more numerous and massive than the luminous ones which the telescope reveals. If the latter, viz. that the star has a velocity which never could have been produced by attraction, "then," as says Professor Newcomb, "it must have been flying forward through space from the beginning, and, having come from an infinite distance, must now be passing through our system for the first and only time." The probability is, however, that the star derived its motion from the source from which it derived its light and heat ; namely, from the collision of the two masses out of which it arose. If the star is ever to be arrested in its onward course, it must be by collision ; but such an event would be its final end.

There are other stars, such as 61 Cygni, ε Indi, Lalande 21258, Lalande 21185, μ Cassiopeiæ, and Arcturus, possessed of motions which could not have been derived from gravity. And there are probably many more of which, owing to their enormous distances, the proper motions have not been detected. a Centauri, the nearest star in the heavens, by less than one-half, is distant twenty-one millions of millions of miles ; and there are, doubtless, many visible stars a thousand times more remote. A star at this distance, though moving transversely to the observer at the enormous rate of 100 miles per second, would take upwards of thirty years to change its position so much as one second, and consequently 1,800 years to change its position one minute. In fact, we should have to watch the star for a generation or two before we could be certain whether it was moving or not.

IV. *Probable Origin of Comets.*

Great difficulty has been experienced in accounting for the origin of comets upon the nebular hypothesis. They approach the sun from all directions, and their motions, in relation to the planets, are as often retrograde as direct. Not only are their orbits excessively elliptical, but they are also inclined to the ecliptic at all angles from 0° to 90°. It is evidently impossible to account satisfactorily for the origin of comets if we assume them all to have been evolved out of the solar nebula, although this has been attempted by M. Faye and others. Comets are evidently, as Laplace and Professor A. Winchell both conclude, strangers to our system, and have come from distant regions of space. If they belonged to the solar system they could not, says Professor Winchell, have parabolic and hyperbolic paths. "Only a small portion of the comets," he remarks, "are known to move in elliptic orbits." [1] This assumption that they are foreigners will account for all the peculiarities of their motions ; but how are we to account for their coming into our system ? How did they manage to leave that system in which they had their origin ? If a comet have come from one of the fixed stars trillions of miles distant, the motion by which it traversed the intervenient space could not, possibly, have been derived from gravity. We are therefore obliged to assume that the motion was a projected motion. Comets, in all probability, have the same

[1] *World Life*, p. 27.

origin as meteorites. The materials composing them, like those of the meteorites, were probably projected from nebulæ by the expulsive force of the heat of concussion which produced the nebulæ. Some of them, especially those with elliptic orbits, may have possibly been projected from the solar nebula.

V. *Nebulæ.*

It is a curious circumstance that the theory here advanced seems to afford a rational explanation of almost every peculiarity of nebulæ, as I have, on former occasions, endeavoured, at some length, to prove.[1]

1. *Origin of nebulæ.*—We have already seen that the theory affords a rational account of the origin of nebulæ.

2. *How nebulæ occupy so much space.*—It accounts for the enormous *space* occupied by nebulæ. It may be objected that, enormous as would be the original temperature of the solar system produced by the primeval collision, it would nevertheless be insufficient to expand the mass, against gravity, to such an extent that it would occupy the entire space included within the orbit of Neptune. But it will be perceived, from what has already been stated regarding the dispersion of the materials before they had sufficient time to assume the gaseous condition, that this dispersion was the main cause of the gaseous nebula coming to occupy so much space. And, to go

[1] *Philosophical Magazine*, July 1878; *Climate and Cosmology*, chap. xix.

farther back, it was the suddenness and almost in-
stantaneity with which the mass would receive the
entire store of energy, before it had time to assume
even the molten, not to say the gaseous, condition,
which led to tremendous explosions, followed by a
wide dispersion of materials.

3. *Why nebulæ are of such varied shapes.*—
Although the dispersion of the materials would be in
all directions, it would, according to the law of probability, very rarely take place uniformly in every
direction. There would generally be a greater
amount of dispersion in some directions than in
others, and the materials would thus be carried along
various lines and to diverse distances ; and, although
gravity would tend to bring the widely scattered
materials ultimately together into one or more
spherical masses, yet, owing to the exceedingly rarified condition of the gaseous mass, the nebulæ would
change form but slowly.

4. *Broken fragments in a gaseous mass of an excessively high temperature the first stage of a nebula.*—
From what has already been shown, it will be seen
that after the colliding of the two dark bodies the
first condition of the resulting nebula would be an
enormous space occupied by broken fragments of
all sizes dashing against each other with tremendous
velocities, like the molecules in a perfect gas. All
the interspaces between those fragments would be
entirely filled with a gaseous mass, which, at its
earliest stages at least, as in the case of the solar
nebula, would have a temperature probably more than

c 2

one hundred thousand times that of the voltaic arc. Whether such a mass would be visible is a point which can hardly be determined, as we can have no experience on earth of a gas at such a temperature.

That there are some of the nebulæ which appear to consist of solid matter interspersed in a gaseous mass is shown by the researches of Mr. Lockyer [1] and others. In fact, the theory is held by Professor Tait [2] that nebulæ consist of clouds of stones—or meteor-swarms, as Mr. Lockyer would term them— in an atmosphere of hydrogen, each stone of which, moving about and coming into collision with some other, is thereby generating heat which renders the circumambient gas incandescent. In reference to this theory of Professor Tait, Mr. Lockyer says that the phenomena of the spectroscope can be quite well explained " on the assumption of a cloud of stones, providing always that you could at the same time show reasonable cause why these clouds of stones were ' banging about ' in an atmosphere of hydrogen." [3] The theory, however, does not appear to afford any rational explanation of this banging about of the stones to and fro in all directions ; for, according to it, the only force available is gravitation, and this can produce merely a motion of the materials towards the centre of the mass. Under these conditions very little impinging of the stones against each other would take place. But, according to the

[1] *Proceedings of Royal Society*, vol. xliii. p. 117.
[2] *Good Words* for 1875, p. 861.
[3] *Manchester Science Lectures.*

theory here adopted, we have an agency incalculably more effective than gravity, one which accounts not merely for the impact of the stones, but for their very existence as such, inasmuch as it explains both what they are and whence they came.

Mr. Lockyer has recently fully adopted Professor Tait's suggestion as to the nature and origin of nebulæ, and has endeavoured to give it further development. He considers the nebulæ to be composed of sparse meteorites, the collisions of which give the nebulæ their temperature and luminosity. He divides the nebulæ into three groups, " according as the formative action seems working towards a centre ; round a centre in a plane, or nearly so ; or in one direction only." As a result we have globular, spheroidal, and cometic nebulæ.

Globular nebulæ he accounts for in the following manner. " If we," he says, " for the sake of the greatest simplicity consider a swarm of meteorites at rest, and then assume that others from without approach it from all directions, their previous paths being deflected, the question arises whether there will not be at some distance from the centre of the swarm a region in which collisions will be most valid. If we can answer this question in the affirmative, it will follow that some of the meteorites arrested here will begin to move in almost circular orbits round the common centre of gravity.

" The major axes of these orbits may be assumed to be not very diverse, and we may further assume that, to begin with, one set will preponderate over the rest.

Their elliptic paths may throw the periastron passage to a considerable distance from the common centre of gravity ; and if we assume that the meteorites with this common mean distance are moving in all planes, and that some are direct and some retrograde, there will be a shell in which more collisions will take place than elsewhere. *Now, this collision surface will be practically the only thing visible, and will present to us the exact and hitherto unexplained appearance of a planetary nebula—a body of the same intensity of luminosity at its edge and centre—thus putting on an almost phosphorescent appearance.*

" If the collision region has any great thickness, the centre should appear dimmer than the portion nearer the edge.

" Such a collision surface, as I use the term, is presented to us during a meteoric display by the upper part of our atmosphere." [1]

Spheroidal nebulæ, he considers, are produced by the rotation of what was at first a globular rotating swarm of meteorites.

Cometic nebulæ are explained, he considers, " on the supposition that we have either a very condensed swarm moving at a very high velocity through a sheet of meteorites at rest, or the swarm at rest surrounded by a sheet, all moving in the same direction."

In an able and interesting work, which seems almost utterly unknown in England,[2] Professor

[1] *Proc. of Royal Society,* vol. xliv. p. 5.
[2] *World Life, or Comparative Geology,* by Alexander Winchell, LL.D., Professor of Geology and Palæontology in the University of Michigan. Chicago : S. C. Griggs & Co. 1883.

Winchell has advanced views similar to those of Tait and Lockyer regarding the nature and origin of nebulæ. But he, in addition, discusses the further question of the origin of those swarms. I shall have occasion to refer to Professor Winchell's views more fully when we come to the consideration of the pre-nebular condition of the universe.

Amongst the first to advance the meteoric hypothesis of the origin and formation of the solar system was probably the late Mr. Richard A. Proctor. This was done in his work, " Other Worlds than Ours," published in 1870. " Under the continual rain of meteoric matter," he says, " it may be said that the earth, sun, and planets are *growing.* Now, the idea obviously suggests itself that the whole growth of the solar system, from its primal condition to its present state, may have been due to processes resembling those which we now see taking place within its bounds." He further adds : " It seems to me that not only has this general view of the mode in which our system has reached its present state a greater support from what is now actually going on than the nebular hypothesis of Laplace, but that it serves to account in a far more satisfactory manner for the principal peculiarities of the solar system. I might, indeed, go farther, and say that where those peculiarities seem to oppose themselves to Laplace's theory they give support to those I have put forward." [1] He then goes on to show the points wherein his theory seems to him to offer a

---

[1] *Other Worlds*, chap. ix.

better explanation of those peculiarities than that of Laplace.

5. *The gaseous condition the second stage of a nebula.*—The second stage obviously follows as a necessary consequence from the first ; for the fragments, in the case under consideration, possess energy in the form of motion, which, with the heat of their circumambient vapour, is more than sufficient not only to convert the fragments into the gaseous state, but to produce complete dissociation of the chemical elements. The complete transformation of the first stage into the second must, therefore, be simply a matter of time.

According to the laws of probability it may, however, sometimes happen that the two original dark bodies will not collide with force sufficient to confer on the broken fragments the energy required to convert them all into the gaseous condition. The result in this case would, no doubt, be that the untransformed fragments, drawn together by their mutual attractions, would collide and form an imperfect star or sun, without a planet. Such a star might continue luminous for a few thousands or perhaps a few millions of years, as the case might be, when it would begin to fade, and finally disappear. We have here an imperfect nebula, resulting in an imperfect star. In short, we should have in those stellar masses, on a grand scale, what we witness every day around us in organic nature, viz. imperfect formations. Such occasional imperfections give variety and add perfection to the whole. How dreary and monotonous would nature be, were every blade of grass, every plant, every animal, and every face we met formed after the most perfect model !

6. *The gaseous condition essential to the nebular hypothesis.*— It is found that the density of the interior planets of our solar system compared with that of the more remote is about as five to one. The obvious conclusion is that there is a preponderance of the metallic elements in the interior planets and of metalloids in the exterior. It thus becomes evident, as Mr. Lockyer has so clearly shown,[1] that when our solar system existed in a nebulous condition the metallic or denser elements would occupy the interior portion of the nebula and the metalloids the exterior. Taking a section of this nebula from its centre to its circumference, the elements would in the main be found arranged according to their densities : the densest at the centre, and the least dense at the circumference. If we compare the planets with their satellites, we find the same law holding true. The satellites of Jupiter, for example, have a density of about only one-fifth of that of the planet, or about one twenty-fifth of that of our earth, showing that when the planet was rotating as a nebulous mass the more dense elements were in the central parts and the less dense at the outer rim, where the satellites were being formed. Again, if we take the case of our globe, we find, as Mr. Lockyer remarks, the same distribution of materials, proving that when the earth was in the nebulous state the metallic elements chiefly occupied

[1] *Manchester Science Lectures.*

the central regions, and the metalloids those outer
parts which now constitute the earth's crust.

All these facts show that the *sifting* and *sorting*
of the chemical elements according to their densities
must have taken place when our solar system was in
the condition of a nebula. But, further, it seems impossible that this could have taken place had the
materials composing the nebula been in the solid
form, even supposing that they had taken the form of
clouds of stones.

It is equally impossible that the nebula could have
been in the fluid or liquid state during this process.
This is obvious, for the nebula must then have
occupied, at least, the entire space within the orbit of
the most remote planet. But our solar system in the
liquid condition could not occupy one-millionth part
of that space. It is therefore evident that the nebula
must have been in the state of a gas, and a gas of
extreme tenuity.

7. *The mass must have possessed an excessive
temperature.*—There is ample evidence, Mr. Lockyer
thinks, to show that the temperature of the solar
nebula was as great as that of the sun at the present
time. But I think it is extremely probable that, in
some of its stages, the nebula had a very much higher
temperature than that now possessed by the sun.
There must, during the sifting period, have been
complete chemical dissociation, so as to keep the
metals and the metalloids uncombined, and thus
allow the elements to arrange themselves according
to their densities. The nebula hypothesis, remarks

Mr. Lockyer, " is almost worthless unless we assume very high temperatures, because, unless you have heat enough to get perfect dissociation, you will not have that sorting out which always seems to follow the same law."

8. *Gravitation could, under no possible condition, have generated the amount of heat required by the nebular hypothesis.*—The nebular hypothesis does not profess to account for the origin of nebulæ. It starts with matter existing in space in the nebulous condition, and explains how, by condensation, suns, planets &c. are formed out of it. In fact, it begins at the middle of a process : it begins with this fine, attenuated material in the process of being drawn together and condensed under the influence of attraction, and professes to explain how, as the process goes on, a solar system necessarily results. To simplify our inquiry we shall confine our attention to the solar nebula, and consider in the first place how far condensation may be regarded as a sufficient source of heat.

A. *Condensation.*— The heat which our nebula could have derived from condensation up to the time that Neptune was detached from the mass, no matter how far the outer circumference of the mass may have originally extended beyond the orbit of that planet, could not have amounted to over $\frac{1}{7,000,000}$ of a thermal unit (772 foot-pounds) for each cubic foot. It is perfectly obvious that this amount could not have produced the dissociation required ; and without the required dissociation Neptune could never have been formed. Further, it is physically impossible that the materials of which our solar system are composed

could have existed in the gaseous state in a cool condition prior to condensation. Unless possessed of
great heat, even hydrogen could not exist in stellar
space in the gaseous form ; and far less could carbon,
iron, platinum, &c. Before Neptune could have been
formed the whole of the materials of the system must
have possessed heat, not only sufficient to reduce them
to the gaseous state, but sufficient to produce complete dissociation. But by no conceivable means could
gravitation have conferred this amount of heat by the
time that the mass had condensed to just within the
limits of the orbit of Neptune.

B. *Solid globes colliding under the influence of
gravity alone.*—As we have already seen, the view has
been adopted by Sir W. Thomson that the solar
nebula may have resulted from the colliding of cold,
solid globes with the velocity due to their mutual
gravitation alone. He states his views as follows :

" Suppose, now, that 29,000,000 cold, solid globes,
each of about the same mass as the moon, and
amounting in all to a total mass equal to the sun's,
are scattered as uniformly as possible on a spherical
surface of radius equal to one hundred times the
adius of the earth's orbit, and that they are left absolutely at rest in that position. They will all commence falling towards the centre of the sphere, and
will meet there in 250 years, and every one of the
29,000,000 globes will then, in the course of half an
hour, be melted, and raised to a temperature of a few
hundred thousand or a million degrees Centigrade.
The fluid mass thus formed will, by this prodigious

heat, be exploded outwards in vapour or gas all round. Its boundary will reach to a distance considerably less than one hundred times the radius of the earth's orbit on first flying out to its extreme limit. A diminishing series of out-and-in oscillations will follow, and the incandescent globe, thus contracting and expanding alternately, in the course, it may be, of 300 or 400 years, will settle to a radius of forty times the radius of the earth's orbit." [1]

The reason which he assigns for the incandescent globe settling down at a radius forty times that of the earth's orbit is as follows : " The radius of a steady globular gaseous nebula of any homogeneous gas is 40 per cent. of the radius of the spherical surface from which its ingredients must fall to their actual positions in the nebula to have the same kinetic energy as the nebula has."

If the solar nebula thus produced would be swelled out into a spherical incandescent mass with a radius 40 times the radius of the earth's orbit, simply because the globes fell from a distance of 100 times the radius of that orbit, then for a similar reason the mass would have a radius of 400 times that of the earth's orbit had the globes fallen from a distance of 1,000 times the radius, and 400,000 times if the globes had fallen from a distance of 1,000,000 times the radius, and two-fifths of any conceivable distance from which they may have fallen.

Supposing all this to be *physically* possible, which it undoubtedly is not, still the heat generated would not

[1] *Proceedings of Royal Institution*, vol. xii. p. 16.

be sufficient; for, whatever the radius of the nebula might be, its entire energy, both kinetic and potential, is simply what is obtained from gravitation, and this, as we have seen, is insufficient.

9. *Condensation the third and last condition of a nebula.*—According to the gravitation theory, condensation is the first stage of a nebula as well as the last; for, according to it, gravity is the force which both collects together the scattered materials and gives them their heat.[1] Before condensation begins there can, according to the gravitation theory, be no such thing as a nebula properly so called. The materials exist, of course, but they do not exist in the form of a nebula. According to the impact theory which I here advocate, condensation cannot begin till after the nebula has begun to lose the heat with which it was originally endowed.

10. *How nebulæ emit such feeble light.*—The light of nebulæ is mainly derived from glowing hydrogen and nitrogen in a condition of extreme gaseous tenuity; and it is well known that these gases are exceedingly bad radiators. The oxyhydrogen flame, although its temperature is surpassed only by that of the voltaic arc, gives a light so feeble as to be scarcely visible in daylight. The small luminosity of nebulæ is, however, mainly due to a different cause. The

---

[1] Laplace held a more accurate view of the primitive condition of the solar nebula. He considered that, owing to intense heat, the solar mass became expanded to the limits of the remotest planetary orbit of our system; that, in cooling, it began slowly to condense; and that, as condensation went on, planet after planet became detached from the mass. Laplace, however, offered no explanation of the manner in which the primitive nebula obtained its heat.

enormous space occupied by those bodies is not so much due to the heat which they possess as to the fact that their materials were dispersed into space before they had time to pass into the gaseous condition; so that, by the time that this latter state was assumed, the space occupied was far greater than was demanded either by the temperature or by the amount of heat which they originally received. If we adopt the nebular hypothesis of the origin of our solar system, we must assume that our sun's mass, when in the condition of a nebula, extended beyond the orbit of the planet Neptune, and consequently filled the entire space included within that orbit. Even supposing Neptune's orbit to have been its outer limit, which, obviously, was not the case, it would nevertheless have occupied 274,000,000,000 times the space it does at present. We shall assume, as before, that 50,000,000 years' heat was generated by the concussion. Of course, there might have been twice or even ten times that quantity; but it is of no importance what amount is in the meantime adopted. Enormous as 50,000,000 years' heat is, it yet gives, as we shall presently see, only 32 foot-pounds of energy for each cubic foot. The amount of heat due to concussion being equal, as before stated, to 100,000,000,000 foot-pounds for each pound of the mass, and a cubic foot of the sun at his present density of 1·43 weighing 89 pounds, each cubic foot must have possessed 8,900,000,000,000 foot-pounds. But when the mass was expanded sufficiently to occupy 274,000,000,000 times its original space

(which it would do when it extended to the orbit of
Neptune), the heat possessed by each cubic foot
would then amount to only 32 foot-pounds.

In point of fact it would not even amount to so
much, for a quantity equal to upwards of 20,000,000
years' heat would necessarily be consumed in work
against gravity in the expansion of the mass, all of
which would, of course, be given back in the form of
heat as the mass contracted. During the nebulous
condition, however, this quantity would exist in an
entirely different form, so that only 19 out of the
32 foot-pounds per cubic foot generated by concussion
would then exist as heat. The density of the nebula
would be only $\frac{1}{16,248,160}$ that of hydrogen at ordinary
temperature and pressure. The 19 foot-pounds of
heat in each cubic foot would thus be sufficient to
maintain an excessive temperature ; for there would
be in each cubic foot only $\frac{1}{440,000}$ of a grain of matter.
But, although the *temperature* would be excessive, the
quantity both of light and heat in each cubic foot
would of necessity be small. The heat being only $\frac{1}{71}$ of
a thermal unit, the light emitted would certainly be
exceedingly feeble, resembling very much the electric
light in a vacuum-tube.

VI. *Binary Systems.*

The theory affords a rational explanation of the
origin of binary stars. Binary stars, in so far as
regards their motion, follow also, of course, as a consequence, from the gravitation theory. If two bodies
come into grazing collision, " they will," says Sir

William Thomson, " commence revolving round their common centre of inertia in long elliptic orbits. Tidal interaction between them will diminish the eccentricities of their orbits, and, if continued long enough, will cause them to revolve in circular orbits round their centre of inertia." [1] This conclusion was pointed out many years ago by Dr. Johnstone Stoney.

VII. *Sudden Outbursts of Stars.*

The case of a star suddenly blazing forth and then fading away, such as that observed by Tycho Brahe in 1572, may be accounted for by supposing that the star had been struck by one of the dark bodies—an event not at all impossible, or even improbable. In some cases of sudden outbursts, such as that of Nova Cygni, for example, the phenomenon may result from the star encountering a swarm of meteorites. The difficulty in the case of Nova Cygni is to account for the very sudden decline of its brilliancy. This might, however, be explained by supposing that the outburst of luminosity was due to the destruction of the meteorites, and not to any great increase of heat produced in the star itself. A swarm of meteorites converted into incandescent vapour would not be long in losing its brilliancy.

Mr. Lockyer thinks that the outburst was produced by the collision of two swarms of meteorites, and not by the collision of the meteorites with a previously existing star.[2]

---

[1] *Proceedings of Royal Institution*, vol. xii. p. 15.
[2] *Proceedings of the Royal Society*, vol. xliii. p. 140.

Amongst the millions of stars occupying stellar space catastrophes of this sort may, according to the theory, be expected sometimes to happen, although, like the collisions which originate stars themselves, they must, doubtless, be events of but rare occurrence.

VIII. *Star Clusters.*

A star cluster will result from an immensely wide-spread nebula breaking up into a host of separate nuclei, each of which becomes a star. The irregular manner in which the materials would, in many cases, be widely distributed through space after collision would prevent a nebula from condensing into a single mass. Subordinate centres of attraction would be established, as was long ago shown by Sir William Herschel in his famous memoir on the formation of stars;[1] and around these the gaseous particles would arrange themselves and gradually condense into separate stars, which would finally assume the condition of a cluster.

IX. *Age of the Sun's Heat: a Crucial Test.*

When we come to the question of the age of the sun's heat, and the length of time during which that orb has illuminated our globe, it becomes a matter of the utmost importance which of the two theories is to be adopted. On the age of the sun's heat rests the whole question of geological time. A mistake here is fundamental. If gravitation be the only source from which the sun derived its heat, then life

[1] *Philosophical Transactions* for 1811.

on the globe cannot possibly date farther back than 20,000,000 years; for under no possible form could gravitation have afforded, at the present rate of radiation, sufficient heat for a longer period. It will not do to state in a loose and general way, as has been frequently done, that the sun may have been supplying our globe with heat at its present rate for 20,000,000 or 100,000,000 years, for gravitation could have done no such thing; a period of 20,000,000, not 100,000,000, years is the lowest which is admissible on that theory. Not even that length of time would be actually available; for this period is founded on Pouillet's estimate of the rate of solar radiation, which has been proved by Langley to be too small, the correct rate being 1·7 times greater. "Thus," as says Sir W. Thomson, "instead of Helmholtz's 20,000,000 years, we have only 12,000,000." But the 12,000,000 years would not in reality be available for plant and animal life; for undoubtedly millions of years would elapse before our globe could become adapted for either flora or fauna. If there is no other source of heat for our system than gravitation, it is doubtful if we can calculate on much more than half that period for the age of life on the earth. Professor Tait probably over-estimates the time when he affirms "that 10,000,000 years is about the utmost that can be allowed, from the physical point of view, for all the changes that have taken place on the earth's surface since vegetable life of the lowest known form was capable of existing there." [1] And this

[1] *Recent Advances in Physical Sciences*, p. 175.

is certainly about all that can ever be expected from gravitation; mathematical computation has demonstrated that it can give no more. The other theory, founded on motion in space—a cause as real as gravitation—labours under no such limitation. According to it, so far at least as regards the store of energy which may have been possessed by the sun, plant and animal life may date back, not to 10,000,000 years, but to a period indefinitely more remote. In fact, there is as yet no known limit to the amount of heat which this cause may have produced; for this depended upon the velocities of the two bodies at the moment prior to collision, and what these velocities were we have no means of knowing. They might have been 500 miles a second, or 5,000 miles a second, for anything which can be shown to the contrary. Of course I by no means affirm that it is as much as 100,000,000 years since life began on our earth; but I certainly do affirm that, in so far as a possible source of the sun's energy is concerned, life may have begun at a period as remote.

PART II.

EVIDENCE IN SUPPORT OF THE THEORY FROM THE AGE OF THE SUN'S HEAT.

TESTIMONY OF GEOLOGY AND BIOLOGY AS TO THE AGE OF THE SUN'S HEAT.

THE question which we have now to consider is—to which of the two theories does geology lend its testimony? Will the length of time which, according to the gravitation theory, can possibly be allotted satisfy the requirements of geology? In short, are the facts of geology reconcilable with the theory? If not, the theory must be abandoned.

Before the period when geologists felt that they were limited to time by physical considerations, the most extravagant opinions prevailed in regard to the length of geological epochs. So long as the physicist continued to state in a loose and general way that the sun might have been supplying our earth with heat at his present rate for the past 100,000,000 years, no very serious difficulty was felt ; but when geologists came to understand that ten or twenty millions of years were all that could be granted to them, the condition of matters was entirely altered. The belief that the mathematical physicist must be right in his

views as to the age of the sun's heat, and that there
is no possibility of a longer period being admitted,
seems at present to be leading geologists towards the
opposite extreme in regard to the length of geological
time. Attempts have been recently made to compress
the geological history of our globe into the narrow
space allotted by the physicist. The attempt is hope-
less, as well as injurious to geological science. What
misleads is not the belief that gravitation could not
possibly afford a supply of heat sufficient for more
than 20,000,000 years, for this is true; it is the
belief that there was no other source of heat than
gravity.

We shall now consider the evidence which geology
seems to afford as to the age of the sun's heat.
Geology is quite competent to render aid on this
point, for the sun's heat must be at least as old as
life on this globe; and the record of the rocks tells us
when this life first appeared. We require, however,
to be able to measure the time which has elapsed since
these records were left. What we want is absolute
time; not relative time. Much has been done by
geologists in regard to relative time; but this can be
of no service to us in our present inquiry. Unfortu-
nately very little trustworthy work has been done in
the way of determining the absolute length of geological periods. Happily, however, great exactness of
measure is not required. A rough approximation to
the truth will suffice for our present purpose. If it
can be shown to be more than fifteen or twenty
millions of years since life first appeared on the earth,

it will as effectually prove that gravitation alone could not have been the source from which the sun derived his heat as if it were shown that that period was a thousand times more remote. All we have to do is simply to assign an *inferior limit* to the age of life on the earth ; and this can be effectually done by means of the methods, imperfect though they be, which we have at command. As the question of geological time is of some importance in relation to our present inquiry, I shall consider it at some length.

Testimony of Geology : method employed.—What has subsequently proved to be a pretty successful method of measuring geological time suggested itself to my mind during the summer of 1865. It then occurred to me that we might obtain a tolerably accurate measure of absolute geological time from the present rate of subaërial denudation, which might be ascertained in the following way : The rate of subaërial denudation must be equal to the rate at which materials are carried off the land into the sea ; and this is measured by the rate at which sediment is carried down by our river systems. *Consequently, in order to determine the present rate of subaërial denudation, we have only to ascertain the quantity of sediment annually carried down by the river systems.* This gives us the time required to remove any given quantity, say one foot, off the face of the country. If we assume the rate to be pretty much the same during past geological ages, we have a means of telling the time that was occupied in removing any known thickness of strata. But as we never can be perfectly certain that the rate is the same in both cases, the results can, of course, be regarded as only approximately true.

Taking the quantity of sediment discharged into the sea annually by the Mississippi river, as determined by Messrs. Brown and Dickson,[1] I found that it amounted to one foot off the face of the country in 1,388 years, and that, at this rate of denudation, our continents, even if they had an elevation of 1,000 feet, would not remain above sea-level over 1,500,000 years.[2] This was an exaggerated estimate of the quantity of sediment, for I shortly afterwards found that far more accurate determinations were made by Messrs. Humphreys and Abbot,[3] who were employed by the United States Government to report upon the physics and hydraulics of the Mississippi. Messrs. Brown and Dickson had estimated the quantity of sediment at 28,188,083,892 cubic feet, whereas Messrs. Humphreys and Abbot found it to be only 6,724,000,000 cubic feet, or less than one-fourth that amount. This gives one foot in 6,000 years as the rate of denudation.

[1] *Proceedings of the American Association for the Advancement of Science* for 1848.

[2] *Philosophical Magazine*, February 1867. I was not aware at this time that Mr. Alfred Tylor had previously applied the same method to determine an entirely different point, viz.: how much the sea-level is being raised by the sediment deposited on the sea-bottom. Mr. Tylor's paper, entitled " On Changes of the Sea-Level effected by existing Physical Causes during stated Periods of Time," appeared in the *Phil. Mag.* for April 1853. Mr. Tylor came to the conclusion that the sea-level was being raised, from this cause, about 3 inches in 10,000 years.

[3] *Report upon the Physics and Hydraulics of the Mississippi.*

At this time Dr. Archibald Geikie took up the question and went into the consideration of the subject in a most thorough manner ; and it is mainly through the instrumentality of his writings on the matter [1] that the method under consideration has gained such wide-spread acceptance among geologists. After an examination of nearly all that is known regarding the amount of sediment carried down by rivers, he drew up the following table, showing the number of years required by seven rivers to remove one foot of rock from the general surface of their basins.

Danube	.	.	.	.	6,846 years
Mississippi	.	.	.	.	6,000 ,,
Nith .	.	.	.	.	4,723 ,,
Ganges	.	.	.	.	2,358 ,,
Rhone	.	.	.	.	1,528 ,,
Hoang-Ho	.	.	.	.	1,464 ,,
Po .	.	.	.	.	729 ,,
		Mean	.	.	3,378

This gives a mean of 3,378 years to remove one foot, or a little over one-half the time taken by the Mississippi. This mean appears to be generally taken as representing the average rate of subaërial denudation of the whole earth, but it has, I fear, been rather too hastily adopted. To estimate correctly the quantity of sediment annually discharged by a large river is a most difficult and laborious undertaking. A perusal of the voluminous report of Messrs. Humphreys and Abbot, extending over 690 pages, which

[1] *Trans. of Geol. Soc. of Glasgow*, vol. iii.; Jukes & Geikie's *Manual of Geology*, chap. xxv.; *Text Book of Geology*, p. 441.

Dr. Geikie justly styles a model of patient and exhaustive research, will clearly show this, and at the same time prove how skilfully and accurately the task allotted to them was performed.

The risk of making very serious errors in computing the amount of sediment discharged, unless proper precautions are taken, is well illustrated in the case of the determinations made by Messrs. Brown and Dickson, to which reference has already been made. Although their report shows that they took great pains in order to arrive at correct results—in fact, they computed the total annual quantity of sediment discharged to within a cubic foot—after all, instead of being correct to this minute quantity, they gave a total more than fourfold what it ought to be. A somewhat similar discrepancy exists in reference to the denudation of the basin of the Ganges. The time required to lower its surface by one foot is, according to one estimate, 2,358 years; according to another, 1,751; and according to a third, only 1,146 years. The first figure is probably nearest the truth. Still, these differences show both the difficulty of the problem and the necessity of caution in adopting any of these results as correct.

By far the most trustworthy determinations of the whole are those of the Mississippi by Messrs. Humphreys and Abbot, which may be relied upon as not far from the truth. But, supposing the estimates in the foregoing table to be perfectly correct, can we assume that their mean may be safely taken as probably representing the average rate of denudation of the whole

earth? I would most unhesitatingly reply, Certainly
not. The Rhone and Po are full of glacier mud
from the Alps; and the amount of sediment which
they carry down may give us the rate of denudation
of Switzerland, but certainly not that of the whole
earth, or even of Europe. The same may be said of
the Ganges, which is charged with the mud which it
brings down from the Himalaya Mountains. The
Hoang-Ho, or Yellow River, is an exceptionally muddy
river; in fact, it derives its name from the vast quantity of yellow mud held by its waters in a state of
solution. It was probably the exceptionally muddy
character of the Po, the Rhone, the Ganges, and the
Yellow River which attracted attention, and led to
observations being made of the sediment they contain.
Rivers more unsuitable than these to give us the
average denudation of the earth's surface could not
well be selected. Among the seven rivers in the table,
leaving out of account the small Scottish stream, the
Nith, with its basin of only 200 square miles, there are
only two, the Mississippi and the Danube, that drain
countries which may be regarded as in every way
resembling the average condition of the earth's surface. I would choose the Mississippi as being supe-
rior to the Danube, for two reasons: (1) because
the rate of denudation of its basin has been more
accurately determined; and (2) because the area of its
basin not only exceeds that of the Danube as five to
one, but better fulfils the necessary conditions, as Sir
Charles Lyell has so clearly shown. "That river,"
says Sir Charles, "drains a country equal to more

than half the continent of Europe, extends through twenty degrees of latitude, and therefore through regions enjoying a great variety of climate, and some of its tributaries descend from mountains of great height. The Mississippi is also more likely to afford us a fair test of ordinary denudation, because, unlike the St. Lawrence and its tributaries, there are no great lakes in which the fluviatile sediment is thrown down and arrested on its way to the sea." [1] There is no other river in the globe which to my mind better fulfils the required conditions. It is no doubt true that the rate of denudation of the basin of the Mississippi is probably less than that of Switzerland, Norway, and the Himalayas, where glaciers abound, and certainly less than that of Greenland and the Antarctic continent ; but, on the other hand, this rate is certainly much greater than that of the whole continent of Africa, Australia, and large tracts of Asia, where the rainfall is much smaller. One foot in 6,000 years may, therefore, I think, be safely taken as the average rate of denudation of the whole surface of the globe.

The average rate of denudation in the past probably not much greater than in the present.—The belief has long prevailed that the rate of denudation was much greater in past ages than it is now; but I am unable to perceive any good grounds for concluding that such was the case at any time since the beginning of the Palæozoic period. Various reasons have, however, been assigned for this supposed greater rate ;

[1] *Student's Elements of Geology*, p. 91.

and to the consideration of these I shall now very briefly refer.

It has been thought that at some remote epoch of the earth's history, when the moon was much nearer and the day much shorter than now, the rate of denudation would, owing to the erosive power of the enormous tides which would then prevail, be much greater than at the present day. This, however, is very doubtful. There is nothing in the stratified rocks which affords any support to the idea of great tidal waves having swept over the land, at least since the time when life began on our globe. Such a state of things would have destroyed all animal life. "The Palæozoic sediments," as Professor A. Winchell remarks, "have been deposited, for the chief part, in quiet seas. The deep beds of limestones and shales are spread out in sheets continent-wide, which testify unmistakably to placid waters and slow deposition." [1] But high tides, not sweeping over the land, would not increase the rate of denudation to the extent supposed. High tides silt up a river channel more readily than they deepen it. A higher tide would probably produce a greater destruction of sea-coast: it would tend to increase the rate of marine denudation, but this would not materially affect the general rate of denudation. For, as the present rate of marine denudation is to that of subaërial denudation only as 1 to about 1,700,[2] it would take a very large increase in the rate of marine denudation to affect sensibly the general result. Suppose the rate of marine denuda-

[1] *World Life*, p. 265.   [2] See *Climate and Time*, p. 337.

tion to have been, for example, ten times as great during the Palæozoic age as it is now (which it certainly was not), it would only have shortened the time required to effect a given amount of denudation of the whole earth by 9 years in 1,700, i.e. by little more than one-half per cent.

Again, it is assumed that the greater rate of terrestrial rotation in the early ages would produce certain influences which would in turn bring about a greater amount of denudation. The rate of rotation has been slowly decreasing for ages, and in Palæozoic times it must, of course, have been greater than at present. A more rapid rotation would increase the velocity of the trade and anti-trade winds, and would thus tend to augment the action of those meteorological agents chiefly effective in the work of subaërial denudation. Here again the testimony of geology is negative. We have no geological grounds to conclude that the winds of Palæozoic times were stronger than those at the present day. The heat was no doubt greater, and perhaps there was more rain; but, on the other hand, there would be less frost, snow, ice, and other denuding agents.

There is one cause which would, perhaps, be more effective than any of the foregoing : viz. the periodic occurrence of glacial epochs. When a country is buried under ice, the erosion of the surface is great. But it must be borne in mind that the influence of rain, rivers, and other denuding agents now in operation would then, in the glaciated regions, be almost *nil*. Besides, the greater part of the materials ground off the rocks would be left on the land; and, when the ice disappeared, it would be found in the form of a thick mantle of boulder clay—a mantle which would protect the rocky surface of the country for thousands and tens of thousands of years from further denudation. This is shown by the fine striæ on the rocky surface, made perhaps more than 50,000 years ago, remaining under the boulder clay as perfect as the day on which they were engraved. But, more than all this, a very considerable part of the 1 foot presently being removed off the country in 6,000 years consists of the loose materials belonging to the glacial epoch, such as sands, gravels, and boulder clay, which are being swept off the surface by rain and river action. Were it not for this, the present rate of subaërial denudation would not be so high as it actually is. Taking all things into consideration, it is, I think, obvious that the average rate of denudation since the beginning of Palæozoic times was probably not much greater than at the present day.

How the method has been applied.—Having determined what appears to be the probable average rate of subaërial denudation, we may now proceed to consider the way in which this rate has been applied to measure past geological time. There are two ways in which it may be applied for this purpose. It may (1) be applied directly: knowing the thickness of strata which may have been removed by denudation, we can easily tell, from that rate, the time it required to effect their removal. If we have evidence, for example, that at some epoch 1,000 feet of stratified rock were carried away, then, on the assumption that the rate of denudation was the same at that epoch as now, we have $1,000 \times 6,000 = 6,000,000$ years as the required time. (2) It may be applied indirectly : knowing the thickness of the strata, we may estimate the time required for their formation. This is the way in which it has usually been applied, but, as we shall see, it is the less satisfactory way of the two.

Dr. A. Geikie gives the land area of the globe as 52,000,000 square miles, and that of water as 144,712,000 square miles.[1] We may thus take the proportion of land to water roughly as 1 to 3 ; about one-quarter of the earth's surface being land, and three-quarters water. One foot, therefore, removed off the surface of the land would cover the whole globe with a layer 3 inches thick, or the entire sea-bottom with a layer 4 inches thick.

If we knew the total quantity of stratified rock on the globe, we could easily tell the time that would be required for its formation. Most geologists would, I believe, be inclined to admit that, if spread uniformly over the entire globe, it would form a layer of at least 1,000 feet in thickness. In such a case the time required for its deposition would be as follows :

$$1,000 \times 6,000 \times 4 = 24,000,000 \text{ years.}$$

This would not, however, represent the age of the stratified rocks. It would only represent the time required to deposit the rocks which we have assumed

[1] *Physical Geography*, p. 103.

to be now in existence. The greater mass of sedimentary rocks has been formed out of previously existing sedimentary rocks, and these again out of sedimentary rocks still older. The oldest known sedimentary rocks are the Laurentian; but these are believed by geologists to have been formed from still older sedimentary rocks. It is therefore evident that the materials composing our stratified beds must have passed through many cycles of destruction and re-formation. The materials of some of the recent formations, for example, may have passed through denudation and deposition a dozen of times over.[1] The time required to have deposited at a given rate the present existing mass of sedimentary rocks is probably but a small fraction of the time required to have deposited at the same rate the total mass that has actually been formed. Few geologists, I think, who will duly reflect on the subject, will deem it too much to say that the present existing stratified rocks have on an average passed at least thrice through the cycle of destruction and re-formation. If this be admitted, then the 1,000 feet of stratified rock represent, not a period of 24,000,000 years, but a period three times as great, viz. 72,000,000 years.

It is impossible to tell from geological data the actual age of the stratified rocks; but this is not

---

[1] It is this destruction of the stratified rocks which makes it so difficult to detect the marks of former glacial epochs, and which has led to such prevailing misconceptions regarding the evidence which we ought to expect of those epochs. See paper read before the Geological Society, " On Prevailing Misconceptions regarding the Evidence which we ought to expect of former Glacial Periods," January 23, 1889.

required.   What we require is, as already stated, not their *actual* age, but an *inferior limit* to that age.

Method as applied by Professor Haughton.—Professor Haughton estimates the mass of the stratified rocks down to the time of the Miocene Tertiary period as being 177,200 feet in thickness, and covering an area equal to that of the sea.   The present rate of subaërial denudation he considers to be equal to one foot removed off the surface of the land in 3,090 years. If the proportion of land to water be taken as 52 to 145, it thus requires 8,616 years to deposit one foot of sediment over the bed of the ocean, and consequently this is the rate at which strata are at present being formed.   This would give $8,616 \times 177,200 = 1,526,750,000$ years for the age of the stratified rocks.   But he assumes the rate of denudation to have been *ten times* greater in geological time than at present.   This consequently reduces the age of the rocks to 152,675,000 years.   By adding one-third for the time which has elapsed since the Miocene Tertiary period he gets 200,000,000 years as a minimum length of geological time.[1]

The validity of this result rests upon what appear to me to be two very doubtful assumptions.   It is assumed in his calculations that the total amount of strata formed during past ages (not the amount presently remaining) was equal to a mass 177,200 feet in thickness, covering the entire area of the ocean.   This is certainly doubtful.   It may have been as great, for anything that can be proved to the

[1] *Physical Geography*, p. 94.

contrary; but we have no evidence that it was so. Certainly there is no evidence that the rate of subaërial denudation during past ages was ever ten times as great as it is now. But how is a length of 200,000,000 years to be reconciled with the age of the sun's heat? The stratified rocks may be as old as this, but assuredly they are not if gravitation was the only source from which the sun derived his energy.

Method as applied by Mr. Alfred R. Wallace.— Mr. Wallace adopts Professor Haughton's estimate of 177,200 feet for the maximum thickness of the sedimentary rocks. But, instead of supposing, like Professor Haughton, the products of denudation to be uniformly spread over the entire sea-bottom, he supposes them spread over a belt of merely 30 miles broad, extending along the entire coast-line of the globe, which he estimates at 100,000 miles. This gives an area of 3,000,000 square miles on which the denuded matter of the whole land area of 57,000,000 square miles is deposited. These two areas are to one another as 1 to 19, and thus it follows that deposition goes on 19 times as fast as denudation. The rate of denudation he takes as one foot removed off the surface of the land in 3,000 years, so that the rate of deposition would be about one foot in 158 years, and consequently the time required to deposit the 177,200 feet of rock would be

$$177,200 \times 158 = 27,997,600 \text{ years.}$$

This is a period double what the gravitation theory

of the source of the sun's energy can afford. And if the rate of denudation be taken at one foot in 6,000 years, which is, as we have seen, probably nearer the truth, then this would make the age of the stratified rocks 56,000,000 years.

There seems to be a little ambiguity about Mr. Wallace's result. Do the 177,200 feet represent the quantity of rock which presently exists, or do they represent the total quantity which has been formed during all past ages? If the former, then the 28,000,000 years is but a fraction of the time which must have been required; for, as we have been shown, the materials composing the stratified rocks have, on an average, been deposited at least three or four times over. If, on the other hand, the thickness is meant to represent the total quantity of rock which has been formed during the whole of past geological time, then the question arises, by what means could this quantity possibly be ascertained? In other words, how was the relation between the present quantity and the total quantity ascertained? But in either case the result is wholly irreconcilable with the gravitation theory of the source of the sun's heat.

Method as applied directly.—We have seen that it is impossible to determine the actual age of the earth from the stratified rocks, even if we knew with perfect accuracy their present total amount. We have also seen that from the rate of deposition we cannot fix with any degree of certainty a minimum value for the age of these rocks. We can, however, by means of

the first or direct application of the method, assign
with tolerable accuracy, as was shown on a former
occasion,[1] a minimum age to the earth. We can be
far more certain of the time which must have been
required to remove by denudation, say, a thousand
feet of rock than we can possibly be of the time required to have deposited a thousand feet of sediment.
The thousand feet of sediment may, under certain
conditions, have been deposited in a hundred years,
while under other conditions they may have required
a million of years. In fact, nothing can be more un-
certain than the rate of deposition : it depends upon
such a multitude of circumstances. At the mouth of
a great river, for example, a foot of sediment may be
deposited in a single day, whereas in some places, as
in mid-ocean, it may require a million of years to deposit the same amount. But in reference to subaërial
denudation no such uncertainty exists.

The utter inadequacy of a period of 20,000,000
years for the age of our earth is demonstrable from
the enormous thickness of rock which is known to
have been removed off certain areas by denudation.
I shall now briefly refer to a few of the many facts
which might be adduced on this point.

Evidence from "faults."—One plain and obvious
method of showing the great extent to which the
general surface of the country has been lowered y
denudation is furnished, as is well known, by the way
in which the inequalities of surface produced by faults

---

[1] *Quart. Journ. of Science*, July 1877 ; *Climate and Cosmology*,
chap. xvii.

or dislocations have been effaced. It is quite common to meet with faults where the strata on the one side have been depressed several hundreds—and in some cases thousands—of feet below those on the other; but we seldom find any indications of such on the surface, the inequalities on the surface having been all removed by denudation. Now, in order to effect this, a mass of rock must have been removed equal in thickness to the extent of the dislocation. The following are a few examples of large faults :

The great Irwell fault, described by Professor Hull,[1] which stretches from the Mersey west of Stockport to the north of Bolton, has a throw of upwards of 3,000 feet.

Some remarkable faults have been found by Professor Ramsay in North Wales. For example, near Snowdon, and about a mile E.S.E. of Beddgelert, there is a fault with a downthrow of 5,000 feet ; and in the Berwyn Hills, between Bryn-mawr and Post-gwyn, there is one of 5,000 feet. In the Aran Range there is a great fault, designated the Bala fault, with a downthrow of 7,000 feet. Again, between Aran Mowddwy and Careg Aderyn the displacement of the strata amounts to no less than from 10,000 to 11,000 feet.[2] Here we have evidence that a mass of rock, varying from one to two miles in vertical thickness, must have been denuded in many places from the surface of the country in North Wales.

The fault which passes along the east side of the

---

[1] *Mem. Geol. Survey of Lancashire*, 1862.
[2] *Mem. Geol. Survey of Great Britain*, vol. iii

Pentlands is estimated to have a throw of upwards of 3,000 feet.[1] Along the flank of the Grampians a great fault runs from the North Sea at Stonehaven to the estuary of the Clyde, throwing the Old Red Sandstone on end sometimes for a distance of two miles from the line of dislocation. The amount of the displacement, Dr. A. Geikie[2] concludes, must in some places be not less than 5,000 feet, as indicated by the position of occasional outliers of conglomerate on the Highland side of the fault.

The great fault crossing Scotland from near Dunbar to the Ayrshire coast, which separates the Silurians of the South of Scotland from the Old Red Sandstone and Carboniferous tracts of the North, has been found by Mr. B. N. Peach, of the Geological Survey,[3] to have in some places a throw of fully 15,000 feet. This great dislocation is older than the Carboniferous period, as is shown by the entire absence of any Old Red Sandstone on the south side of the fault, and by the occurrence of the Carboniferous Limestone and Coal-measures lying directly on the Silurian rocks. We obtain here some idea of the enormous amount of denudation which must have taken place during a comparatively limited geological epoch. So vast a thickness of Old Red Sandstone could not, as Mr. Peach remarks, " have ended originally where the fault now is, but must have swept southwards over the Lower Silurian uplands. Yet these thousands of

---

[1] *Memoir* to Sheet 32, Geol. Survey Map of Scotland.

[2] *Nature*, vol. xiii. p. 390.

[3] *Explanation* to Sheet 15, Geol. Survey Map of Scotland.

feet of sandstones, conglomerates, lavas, and tuffs were so completely removed from the south side of the fault previous to the deposition of the Carboniferous Limestone series and the Coal-measures, that not a fragment of them is anywhere to be seen between these latter formations and the old Silurian floor."[1] This enormous thickness of nearly three miles of Old Red Sandstone must have been carried away during the period which intervened between the deposition of the lower members of the Lower Old Red Sandstone and the accumulation of the Carboniferous Limestone.

Near Tipperary, in the south of Ireland, there is a dislocation of the strata of not less than 4,000 feet,[2] which brings down the Coal-measures against the Silurian rocks. Here 1,000 feet of Old Red Sandstone, 3,000 feet of Carboniferous Limestone, and 800 feet of Coal-measures have been removed by denudation off the Silurian rocks. Not only has this immense thickness of beds been carried away, but the Silurian itself on which they rested has been eaten down in some places into deep valleys several hundreds of feet below the surface on which the Old Red Sandstone rested.

[1] I have been informed by Mr. Peach that since the above was written additional light has been cast on this immense fault. It has been found, he says, that the fault consists of two sub-parallel branches, the more southerly of which has the effect of bringing the rocks of the Upper Silurian age against the Lower Silurian beds The northern branch brings the upper division of the Lower Old Red Sandstones, in turn, against the Upper Silurian rocks. This, Mr. Peach remarks, does not in the least invalidate the reasoning as to the amount of material removed by denudation from this region in the time specified. In fact, it shows, he says, that a greater amount must have been removed than was at first suspected.

[2] Jukes's and Geikie's *Manual of Geology*, p. 441.

Turning to the American continent, we find the amount of rock removed to be even still greater. In the Valley of Thessolon, to the north of Lake Huron, there is a dislocation of the strata to the extent of 9,000 feet.[1]

In front of the Chilowee Mountains there is a vertical displacement of the strata of more than 10,000 feet.[2] Professor H. D. Rogers found in the Appalachian coal-fields faults ranging from 5,000 feet to more than 10,000 feet of displacement.

In the Nova Scotia coal-fields one or two miles in thickness of strata have been removed in some places.[3]

A great fracture runs along the axis of the Sierra Nevada for 300 miles, accompanied by a dislocation of 3,000 to 10,000 feet.[4]

The anticlinal of the Park Range of the Rocky Mountains was cleft down the axis, and the eastern half depressed 10,000 feet. And Mr. J. P. Lesley gives an account of a fault in the Appalachians of not less than 20,000 feet, bringing the upper Devonian strata on the one side opposite to the lowest Cambrian on the other.[5]

A fault with a vertical displacement of 20,000 feet was found in the Uinta Mountains.[6]

In the Aqui range of mountains, Utah, there is a

[1] *Geology of Canada*, 1863, p. 61.
[2] Safford's *Geology of Tennessee*, p. 309.
[3] Lyell's *Student's Manual*, chap. xxiii.
[4] *Geological Studies*, by Prof. A. Winchell, p. 165.
[5] *Geological Studies*, pp. 93, 163.
[6] Powell's *Geology of the Uinta Mountains*.

fault determined by Mr. S. F. Emmons to be at least
10,000 feet.[1]

The Grand Cañon of Colorado, in some places
4,000, 5,000, and 6,000 feet in depth, is cut, says Professor A. Winchell, in a plateau which has itself been
lowered by erosion to the extent of 10,000 feet; and
this plateau occupies an area of 13,000 to 15,000
square miles.[2]

The Grand "Wash Fault," Colorado, has a downthrow to the west of 6,000 feet. The "Hurricane
Fault," close to it, has displaced the strata to the
extent of over 12,000 feet.[3]

In the Valley of East Tennessee, Appalachian
Mountains, it has been shown by Mr. J. P. Lesley
that as much as 35,000 feet of rock have been removed by denudation. But this being from an anticlinal arch, it does not, of course, afford any measure
of the extent of the denudation of the surrounding
country. Major J. W. Powell, Director of the U.S.
Geol. Survey, found that under a similar condition
as much as three and a half miles of strata have been
removed by denudation from the top of anticlinal beds
in the Uinta Mountains.[4]

Probably the most enormous displacement of
strata which has yet been found is that of the
"Wahsatch Fault," Utah. This fault is about 100

[1] *Geological Exploration of the Fortieth Parallel,* vol. ii. p. 456.
[2] *Geological Studies,* p. 92; see also Dutton's *Tertiary History of the Cañon District.*
[3] *Tertiary History of the Cañon District,* pp. 20, 113; *Second Annual Report, U.S. Geol. Survey,* p. 125.
[4] Powell's *Geology of Uinta Mountains.*

miles in length, crossing the fortieth parallel of latitude from north to south, with a downthrow to the west of not less than 40,000 feet. So clear is the evidence regarding this fault that Mr. Clarence King says "that there can be no doubt of the quantitative correctness of my reading of this tremendous dislocation." [1]

There are other modes than the foregoing by means of which geologists are enabled to measure the thickness of strata which may have been removed in places off the present surface of the country. Into the details of these I need not here enter ; but I may give a few examples of the enormous extent to which the country, in some places, has been found to have been lowered by denudation.

Dr. A. Geikie has shown [2] that the Pentlands must at one time have been covered with Carboniferous rocks, upwards of a mile in thickness, which have all been removed by denudation.

In the Bristol coal-fields, between the river Avon and the Mendips, Sir Andrew C. Ramsay has shown [3] that about 9,000 feet of Carboniferous strata have been removed by denudation from the present surface.

Between Bendrick Rock and Garth Hill, South Glamorganshire, a mass of Carboniferous and Old Red Sandstone, of upwards of 9,000 feet, has been removed. At the Vale of Towy, Caermarthenshire, about 6,000 feet of Silurian and 5,000 feet of Old Red

[1] *Geological Exploration of the Fortieth Parallel*, vol. i. p. 745.
[2] *Memoir* to Sheet 32, Geol. Survey of Scotland.
[3] *Denudation of South Wales. Memoirs of Geol. Survey*, vol. i.

Sandstone—in all about 11,000 vertical feet—have been swept away. Between Llandovery and Aberaeron a mass of about 12,000 vertical feet of the Silurian series has been removed by denudation. Between Ebwy and the Forest of Dean, a distance of upwards of 20 miles, a thickness of rock varying from 5,000 to 10,000 feet has been abstracted.

Prof. Hull found[1] on the northern flanks of the Pendle Range, Lancashire, the Permian beds resting on the denuded edges of the Millstone Grit, and these were again observed resting on the Upper Coal-measures south of the Wigan coal-field. Now from the known thickness of the Carboniferous series in this part of Lancashire he was enabled to calculate approximately the quantity of Carboniferous strata which must have been carried away between the period of the Millstone Grit and the deposition of the Permian beds, and found that it actually amounted to no less than 9,900 feet. He also found in the Vale of Clitheroe, and at the base of the Pendle Range, that the Coal-measures, the whole of the Millstone Grit, the Yoredale series, and part of the Carboniferous Limestone, amounting in all to nearly 20,000 feet, had been swept away—an amount of denudation which, as Prof. Hull remarks, cannot fail to impress us with some idea of the prodigious lapse of time necessary for its accomplishment.

It may be observed that, enormous as is the amount of denudation indicated by the foregoing figures, these figures do not represent in most cases

---

[1] *Quart. Journ. Geol. Soc.* vol. xxiv. p. 323.

the actual thickness of rock removed from the surface. We are necessitated to conclude that a mass of rock equal to the thickness stated must have been removed, but we are in most cases left in uncertainty as to the total thickness which has actually been carried away. It cannot be imagined that these great disruptions occurred first when the surface became subject to denuding agencies, or that denudation ceased to operate precisely when the inequality was smoothed away. Moreover, during the time the surface on one side of the fault was being reduced, some amount of denudation must also have been in progress on the other and lower side. In the case of a fault, for example, with a displacement of, say, one mile, where no indication of it is seen at the surface of the ground, we know that on one side of the fault a thickness of rock equal to one mile must have been denuded, but we do not know how much more than that may have been removed. For anything which we know to the contrary, hundreds of feet of rock may have been removed before the dislocation took place, and as many more hundreds after all indications of dislocation had been effaced at the surface.

But it must be observed that the total quantity of rock which has been removed from the *present* surface of the land is evidently small in proportion to the total quantity removed during the past history of our globe. For those thousands and thousands of feet of rock which have been denuded were formed out of the waste of previously existing rocks, just as these had been formed out of the waste of yet older rock-masses.

In short, as a general rule, the rocks of one epoch have been formed out of those of preceding periods, and go themselves to form those of subsequent epochs.

In many of the cases of enormous denudation to which we have referred, the erosion has been effected during a limited geological epoch. We have, for example, seen that upwards of a mile in thickness of Carboniferous rock has been denuded in the area of the Pentlands. But the Pentlands themselves, it can be proved, existed as hills, in much their present form, before the Carboniferous rocks were laid down over them; and as they are of Lower Old Red Sandstone age, and have been formed by denudation, they must consequently have been carved out of the solid rock between the period of the Old Red Sandstone and the beginning of the Carboniferous age. This affords us some conception of the immense lapse of time represented by the Middle and Upper Old Red Sandstone periods.

Again, in the case of the great fault separating the Silurians of the south of Scotland from the Old Red Sandstone tracts lying to the north, a thickness of the latter strata of probably more than a mile, as we have seen, must have been removed from the ground to the south of the fault before the commencement of the Carboniferous period. And again, in the case of the Lancashire coal-fields, to which reference has been made, nearly two miles in thickness of strata had been removed in the interval which elapsed between the Millstone Grit and the Permian periods.

Time required to effect the foregoing amount of

denudation.— To lower the country one mile by denudation would therefore require, according to the rate which we have already established, about 15,000,000 years ; but we have seen that a thickness of rock more than equal to that must have been swept away since the Carboniferous period ; and even during the Carboniferous period itself more than a mile in thickness of strata in many places was removed. Again, there can be no doubt whatever that the amount of rock removed during the Old Red Sandstone period was much greater than one mile ; for we know perfectly well that over large tracts of country nearly a mile in thickness of rock was carried away *between the period of the Lower Old Red Sandstone and the Carboniferous epoch.* Further, all geological facts go to show that the time represented by the Lower Old Red Sandstone itself must have been enormous.

Now, three miles of rock removed since the commencement of the Old Red Sandstone period (which, doubtless, is an under-estimate) would give us 45,000,000 years.

Again, going farther back, we find the lapse of time represented by the Silurian period to be even more striking than that of the Old Red Sandstone. The unconformities in the Silurian series indicate that many thousands of feet of these strata were denuded before overlying members of the same great formations were deposited. And again, this immense formation was formed in the ocean by the slow denudation of pre-existing Cambrian continents, just as these had been built up out of the ruins of the still prior Laurentian land. And even here we do not reach the end of the series, for the Laurentians themselves resulted from the denudation, not of the primary rocks of the globe, but of previously existing sedimentary and probably igneous rocks, of which, perhaps, no recognisable portion now remains.

It is the opinion of Mr. Darwin, and also of Mr. Wallace, that the geological time which elapsed anterior to the Cambrian period was as long as the whole interval from that period to the present day. This is an opinion which I suppose is supported by most geologists. But, to err on the safe side, I shall assume that the time which had elapsed prior to the Old Red Sandstone was not greater than the time which has elapsed since that period. Even on this assumption we have at least 90,000,000 years as a minimum duration of geological time.

Age of the earth as determined by the date of the glacial epoch.—Professor A. Winchell, by a most careful examination of the probable relative lengths of geological periods, arrived at the conclusion that the time which elapsed since the beginning of the *glacial epoch* is to the time which has elapsed since the solidification of the earth's surface as 1 to 250.[1] According to the eccentricity theory of the cause of the glacial epoch, that epoch began 240,000 years ago; consequently this makes the time since solidification took place 60,000,000 years, a period which agrees roughly with that deduced from denudation,

[1] *World Life*, p. 369.

and is so far presumptive evidence of the truth of that
theory of the cause of the glacial cold.

Testimony of Biology.—The time required for the
variation and modification of organic forms has, Mr.
Alfred R. Wallace states, been generally considered to
require an even longer series of ages than might
satisfy the demands of physical geology. This is a
point, however, on which I am not qualified to venture
an opinion. I shall simply refer to the views held by
our highest authorities on the subject.

Referring to Professor Huxley's anniversary address to the Geological Society in 1870, where he
shows that almost all the higher forms of life must
have existed during the Palæozoic period, Mr. Wallace
says : " Thus, from the fact that almost the whole of
the Tertiary period has been required to convert the
ancestral Orohippus into the true horse, he, Professor
Huxley, believes that, in order to have time for the
much greater change of the ancestral ungulata into
the two great odd-toed and even-toed divisions (of
which change there is no trace even among the earliest
Eocene mammals), we should require a large portion,
if not the whole, of the Mesozoic or Secondary period.
Another case is furnished by the bats and whales,
both of which strange modifications of the mammalian
type occur perfectly developed in the Eocene formation. What countless ages back must we, then, go for
the origin of these groups, the whales from some ancestral carnivorous animal, and the bats from the
insectivora ! And even then we have to seek for the
common origin of carnivora, insectivora, ungulata,

F

and marsupials at a far earlier period ; so that, on the
lowest estimate, we must place the origin of the mam-
malia very far back in Palæozoic times." [1]

" If the very small differences," says Professor
Huxley,[2] " which are observable between the *Crocodilia*
of the older Mesozoic formations and those of the present day furnish any sort of approximation towards
an estimate of the average rate of change among the
Sauropsida, it is almost appalling to reflect how far
back in Palæozoic times we must go before we can
hope to arrive at that common stock from which the
Crocodilia, *Lacertilia*, *Ornithoscelida*, and *Plesiosauria*,
which had attained so great a development in the
Triassic epoch, must have been derived.

" The *Amphibia* and *Pisces* tell the same story.
There is not a single class of vertebrated animals
which, when it first appears, is represented by ana-
logues of the lowest known members of the same class.
Therefore, if there is any truth in the doctrine of
evolution, every class must be vastly older than the
first record of its appearance upon the surface of the
globe. But if considerations of this kind compel us
to place the origin of vertebrated animals at a period
sufficiently distant from the Upper Silurian, in which
the first Elasmobranchs and Ganoids occur, to allow
of the evolution of such fishes as these from a
vertebrate as simple as the *Amphioxus*, I can only
repeat that it is appalling to speculate upon the
extent to which that origin must have preceded the

[1] *Island Life*, p. 204.
[2] *Quart. Journ. of Geol. Soc.* vol. xxvi. p. 53.

epoch of the first recorded appearance of vertebrate life."

" If the theory be true," says Mr. Darwin, " it is indisputable that before the lowest Cambrian stratum was deposited long periods elapsed—as long as, or probably far longer than, the whole interval from the Cambrian age to the present day ; and that during these vast periods the world swarmed with living creatures." [1]

In referring to the abundant and well-developed fauna of the Cambrian period, Sir Andrew C. Ramsay remarks : [2] " In this earliest known varied life we find no evidence of its having lived near the beginning of the Zoological series. In a broad sense, compared with what must have gone before, both biologically and physically, all the phenomena connected with this old period seem, to my mind, to be quite of a recent description ; and the climates of seas and lands were of the very same kind as those that the world enjoys at the present day—one proof of which, in my opinion, is the existence of great glacial boulder beds in the Lower Silurian strata of Wigtonshire, west of Loch Ryan."

Professor Haeckel remarks that " Darwin's theory, as well as that of Lyell, renders the assumption of immense periods absolutely necessary. If the theory of development be true at all, there must certainly have elapsed immense periods, utterly inconceivable to us."

In reference to the foregoing, Mr. Wallace says : [3]

---

[1] *Origin of Species*, p. 286.
[2] *Proceedings of the Royal Society*, No. 152, 1874, p. 342.
[3] *Island Life*, p. 205.

" These opinions, and the facts on which they are
founded, are so weighty that we can hardly doubt
that, if the time since the Cambrian epoch is correctly
estimated at 200,000,000 of years,[1] the date of the
commencement of life on the earth cannot be much
less than 500,000,000; while it may not improbably
have been longer, because the reaction of the organism
under changes of the environment is believed to have
been less active in low and simple than in high and
complex forms of life, and thus the processes of organic
development may for countless ages have been excessively slow."

I think it must now be perfectly evident that the
facts both of geology and of biology are utterly irreconcilable with the theory that the sun's heat was derived
from the condensation of its mass by gravitation; and
that the mistake in regard to geological time has been
committed by the physicist, and not by the geologist.
The grounds upon which the geologists and the
biologists found the conclusion that it is more than
20 or 30 millions of years since life began on the
earth are far more certain and reliable than the
grounds upon which the physicist concludes that the
period must be less. The only real ground that the
physicist has is that according to the theory which he
holds of the origin of the sun's heat a longer period is
not possible. This might be considered good evidence
were no other theory possible; but there is another
theory, which accords with all the facts, and consequently has a strong presumption in its favour.

---

[1] Of course, Mr. Wallace does not believe that it is actually
200,000,000 years since the Cambrian period.

PART III.

EVIDENCE IN SUPPORT OF THE THEORY FROM THE PRE-NEBULAR CONDITION OF THE UNIVERSE.

THE nebular hypothesis, strictly speaking, is one simply intended to account for the origin of our solar system. "It is," as remarks Professor A. Winchell, " primarily a genetic explanation of the phenomena of the solar system; and accessorily a co-ordination, in a common conception, of the principal phenomena in the stellar and nebular firmament, as far as human vision has been able to penetrate."[1] The theory starts with the assumption that all the materials composing the solar system once existed in a state of extreme tenuity and diffusion, filling far more than the entire space included within the orbit of the most remote planet. It begins with this diffused nebulous mass tending slowly, under the influence of gravitation, towards a state of aggregation. Beyond this point the received nebular hypothesis does not extend.

It will be observed that the theory here begins in the middle of a process. It begins with the assumption of a mass in the act of condensing under the influence of gravity. It offers no explanation of the origin of the mass, or how it came to be in this

[1] *World Life*, p. 196.

attenuated state, or in what condition it existed before the materials began to draw together. These
are, however, inquiries which naturally force themselves on our attention. If the nebular theory be a
true theory of the origin of the solar system, then
this nebulous mass must have had an antecedent
history, and we cannot help feeling the instinctive
desire of tracing the chain of causation farther back.
The mind presses towards an absolute beginning.
It is the goal to which it aspires, and no amount of
failure will ever deter it from renewing its efforts.
Of recent years a considerable amount of attention
has been devoted to inquiries in this direction ; nearly
all of which, it is true, has necessarily been of a specu-
lative and hypothetical character. But hypothesis,
as Mr. Lockyer remarks, is the life-blood of investigation.

The nebular hypothesis is so highly probable as to
have gained almost universal acceptance. In fact, it
contains very little of a hypothetical nature. It is, as
Mr. Mill says, " an example of legitimate reasoning
from a present effect to its past cause, according to
the known laws of that cause." Like the hypothesis
of a luminiferous ether, if it is not a true theory, one
would almost think that it deserves to be so.

There seems no reason why inquiries should stop
at the point where Laplace began. The same line of
reasoning may yet carry us back into the pre-nebular
region, and perhaps with as great a degree of certainty
as it has done in the nebular ; though, no doubt, the
farther back we proceed, the more difficult probably

will the inquiry become. But, be all this as it may, there can be little doubt that the path of investigation is a legitimate one, and also one which is worthy of being traced out.

I shall now briefly refer to some of the leading views which have been expressed in regard to the pre-nebular history of the universe, and shall afterwards consider the additional light which the theory discussed in this volume seems to cast on the subject.

The commonly received opinion is that the nebulæ were formed from ordinary matter existing in a high state of division, and widely diffused through space. The " cosmical dust," as it is called, was the universal "world-stuff" out of which all things were supposed to be formed. It is held that in receding backwards in pre-nebular times, the smaller, more simple, and elementary the materials were. Out of this primitive cosmical dust, or world-stuff, by aggregation, the materials became successively larger and more complex. The theory of the origin of nebulæ, on this principle, has been clearly stated by Professor Winchell, and I here give a brief outline of his views on the subject.

Professor A. Winchell on the pre-nebular condition of matter.—This cosmical dust, or world-stuff, he considers to be scattered promiscuously through boundless space. It is cold and non-luminous, and is acted upon by forces of attraction and probably of repulsion. The material particles, either as atoms or less probably as molecules, are drawn by mutual attraction into groups and swarms. Any central attractive force, as of a sun or planet, by causing the particles to move in converging lines, would cause them to become approximated and ultimately aggregated. Thus both mutual attractions and centric movements would tend to produce aggregations dispersed through space. But in the presence of two or more attractive centres, as in the present constitution of the Cosmos, it is impossible that any mass shall fall directly upon its centre of attraction. Hence motions of rotation will be established in the mass, and also orbital motions of masses about each other. In addition to the mutual attraction of the molecules, the convergence of their paths towards centres of attraction must also tend to the formation of masses and swarms of masses and particles. " We have then," he says, " to picture indefinite space as pervaded by swarms of masses and particles of dark matter. Each mass or particle may, nevertheless, be separated by thousands of miles. It is manifest, therefore, that each mass or particle will eventually dispose itself, under the fixed action of the forces of matter, in some definite order. It is manifest also, from what has been said, that each swarm will have a progressive motion along a path having the essential character of an orbit around some dominant centre of attraction. If, as seems to be the fact, an ethereal medium, or any condition of interplanetary matter, exists in space, it opposes the movements of these swarms by opposing the motion of each constituent mass. But the smaller masses—the particles and molecules—would feel this resistance to the greatest extent. They would therefore fall behind the heavier masses, and would be most deflected toward the attracting centre. The smallest particles would be driven farthest to the rear, and dispersed farthest from the orbit of the train, along the side turned toward the principal attraction. The swarm would present an elongated form, in which the larger and heavier masses would move foremost, and nearest the line of the orbit—that is, near the exterior skirt of the area covered by the general swarm—while the smaller ones would follow, in graduated succession, in a long train which would present a fan-like expansion lying mostly on the inside of the path of the principal masses."

" This, it may be conceived, is the mode of aggregation of these cosmical matters in the depths of space. Of course the attractions which control them are feeble ; their movements are slow, the resistances are relatively inconsiderable, and the elongation of the swarm is correspondingly inconspicuous. What I have described is a tendency which would be present. Sometimes the controlling attraction would be only another cosmical swarm. The two swarms would revolve similarly about their common centre of gravity, while prolonged resistances would cause their slow approximation and final coalescence at the common centre of gravity. Sometimes the controlling attraction would be exerted by a distant sun, around which it would slowly move, continually gathering up additions of matter from the wide fields of space."

" In most cases all controlling attraction would be

feebly felt. These clouds of cosmical dust would float
practically poised in the midst of space, and would
gradually grow by the continued accession of new
matter. Some of them would become aggregates of
large dimensions, and their attraction would be dis-
tinctly felt by other aggregates. There would be a
tendency of such aggregates to approach each other.
They might possibly approach along a straight line;
but more probably some third aggregation, or some
distant sun, would deflect them into orbits about their
common centre of gravity, in which, by prolonged
collisions of cosmical matter, they are brought to
ultimate coalescence with each other. Or some other
attractive disturbance affords such a resultant of
actions as may bring them more directly together.
When these larger aggregations of world-stuff come
together, the result is an aggregation approaching
the dimensions of the Herschellian nebulæ." [1]

In regard to the origin of the heat of the nebulæ,
I am glad to find that Professor Winchell, to a certain
extent, adopts the views which I have so long enter-
tained on the subject. " The thought," he says, " must
already have suggested itself to the reader that the
process of conglomeration affords an explanation of
the intense heat which vaporises its substance, and
causes it to yield a spectrum of bright lines. As the
sudden compression of a portion of atmospheric air
yields heat sufficient to ignite tinder, or fuse and
volatilise a descending meteor-mass, so the precipita-
tion of one planet upon another would liberate suffi-

[1] *World Life*, p. 72.

cient heat to reduce them both to a state of fusion, or even of vapour. Still more must the intensest heat be generated by the impact of two nebulous masses, one or both of which together may embrace more matter than all our planets and the sun combined— as much even as the matter of our entire visible firmament of stars. One experiences a distinct feeling of relief in the discovery of such a possible means of ignition of nebulæ."

Mr. Charles Morris on the pre-nebular condition of matter.—Others again suppose matter to be present everywhere throughout space. This view has been ingeniously advocated by Mr. Charles Morris in an article on " The Matter of Space," which appeared in *Nature*, February 8, 1883. The hypothesis of an ether specially distinct from matter he considers to be a gratuitous assumption, and one of the last surviving relics of eighteenth century science, and, unless it can be proved that highly disintegrated matter is positively incapable of conveying light vibrations, there is no warrant for assigning this duty to a distinct form of substance. But that matter exists in outer space in the same conditions as in planetary atmospheres he thinks is improbable. Its duty as a conveyor of radiant vibrations seems to require a far greater tensity, and its disintegration is probably extreme. Assuming matter throughout the universe— here as condensed spheres, and there in outer space as highly rarified substance—the atmospheric envelopes of the spheres, he considers, will gradually shade off into the excessively rare matter of mid-space.

Matter may exist in countless conditions as to simplicity and complexity, &c., but the base particle he
assumes to be the same under all conditions. In the
spheres there is matter ranging from the simplest
elementary gases, through the mineral compounds of
the solid surface, to the highly compounded organic
molecules. In outer space the variation is in the
opposite direction ; the matter existing there in a
highly disintegrated condition.

Every particle he considers to possess a certain
amount of motor energy in the form of heat. As the
total amount of this energy in the universe remains
unchanged, a particle can only lose energy by trans-
ferring it to others. This heat energy acts, of course,
in opposition to gravity : it tends to repel the particles from each other, while gravity, on the other
hand, tends to draw them together. The former acts
as a centrifugal, the latter as a centripetal energy.
If the heat momentum of the particles be insufficient
to constitute a centrifugal energy equal to the centripetal energy of gravitation, then the material contents
of space will be drawn into the attracting spheres as
atmospheric substance, and outer space, in this case,
will be left destitute of matter. If, on the contrary,
the centrifugal energy of the particles be sufficient to
resist gravitation, then the particles will remain free,
and space will continue to be occupied with matter.
As has been stated, the sum of motor energy in
the universe remaining unchanged, the aggregation of atmospheric substance around any planet
resulting from the loss of motor energy must

cause an increase of motor energy in the particles outside.

The theory seems to dispense with the necessity for assuming a luminiferous ether, for the functions attributed to the ether may, it is thought, be performed by the particles themselves; a view which has been advocated by Euler, Grove, and others. The origin of nebulæ, according to the theory, is accounted for as follows :

" The nebular hypothesis," says Mr. Morris, "holds that the matter now concentrated into suns and planets was once more widely disseminated, so that the substance of each sphere occupied a very considerable extent of space. It even declares that the matter of the solar system was a nebulous cloud, extending far beyond the present limits of that system. From this original condition the existing condition of the spheres has arisen through a continued concentration of matter. But this concentration was constantly opposed by the heat energy of the particles, or, in other words, by their centrifugal momentum. This momentum could only be got rid of by a redistribution of motor energy. If, for illustration, the average momentum of the particles of the nebulæ was just equivalent to their gravitative energy, then a portion of this energy must radiate or be conducted outwards ere the internal particles could be held prisoners by gravitation. The loss of momentum inwardly must be correlated with an increase of momentum outwardly.

" This is a necessary [consequence of the heat

relations of matter. As substance condenses, its capacity for heat decreases and its temperature rises, hence a difference of temperature must constantly have arisen between the denser and the rarer portions of the nebulous mass, and equality of temperature could be restored only by heat radiation. This radiation still continues, and must continue until condensation ceases and the temperatures of the spheres and space become equalised; but this is equivalent to declaring that as the particles of the spheres decrease in heat momentum those of interspheral space increase, and if originally the centrifugal and centripetal energies of matter approached equality they must become unequal, centripetal energy becoming in excess in spheral matter, centrifugal energy in the matter of space. Thus, as a portion of the widely distributed nebulous matter lost its heat, and became permanently fixed in place by gravitative attraction, another portion gained heat, became still more independent of gravity, and assumed a state of greater nebulous diffusion than originally. The condensing spheres only denuded space of a portion of the matter which it formerly held, and left the remainder more thinly distributed than before. The spheres, in their concentration, have emitted, and are emitting, a vast energy of motion. This motor energy yet exists in space as a motion of the particles of matter, which, therefore, press upon each other, or seek to extend their limits, with increasing vigour, so that the elasticity of interspheral matter is constantly increasing."

Sir William R. Grove on the pre-nebular con-

dition of matter.—Amongst the first to advocate the view that ordinary matter is everywhere present in space was Sir William R. Grove. In a lecture delivered at the London Institution as far back as January 1842, he stated that it appeared to him that heat and light, according to the undulatory theory, were the result of the vibrations of ordinary matter itself, and not that of a distinct ethereal fluid. Twenty years afterwards, referring to the views he then advanced, he says: " Although this theory has been considered defective by a philosopher of high repute, I cannot see the force of the arguments by which it has been assailed ; and therefore, for the present, though with diffidence, I still adhere to it." [1]

He adduces a great many facts and forcible arguments in support of his position. He says that "there appears no reason why the atmosphere of the different planets should not be, with reference to each other, in a state of equilibrium. Ether, or the highly attenuated matter existing in the interplanetary space, being an expansion of some or all of these atmospheres, or of the more volatile portions of them, would thus furnish matter for the transmission of the modes of motion which we call light, heat, &c." It is assumed in the theory, of course, that matter must form a universal planum.

Sir William Grove favours the idea that the universe is illimitable in extent, a view held by many eminent thinkers.

[1] *Correlation of Physical Forces*, p. 164 (fifth edition), 1867.

EVOLUTION OF THE CHEMICAL ELEMENTS, AND ITS
RELATIONS TO STELLAR EVOLUTION.

We come now to the consideration of a subject
which has a most important bearing on the question
of stellar evolution, viz. the genesis and dissociation
of the chemical elements. The evolution of one
element from another is, it is true, as yet but a mere
hypothesis, but it is an hypothesis well supported by
a host of facts and considerations, and held by a large
number of our leading chemists and physicists. "The
demonstrated unity of force," says Professor F. W.
Clarke,[1] "leads us by analogy to expect a similar unity
of matter; and the many strange and hitherto unexplained relations between the different elements
tend to encourage our expectations." The hypothesis
throws much light on some obscure points in stellar
evolution. In regard to this, Professor Clarke justly
remarks that "it is plain that the nebular hypothesis
would be doubled in importance, and our views of the
universe greatly expanded, if it could be shown that
an evolution of complex from simple forms of matter
accompanied the development of planets from the
nebulæ. Evolution could look for no grander triumph."
In fact, it is difficult to understand how our sun and
the stars could have been evolved from nebulæ without
assuming an evolution of the chemical elements. The
true nebulæ show the presence of only two elements,
nitrogen and hydrogen, but our sun contains more

[1] *Popular Science Monthly* for January 1873.

than a dozen of distinct elements, and the planets more than three times that number. How, then, could all these have arisen out of nebulæ composed simply of nitrogen and hydrogen ? The matter is plain if we assume an evolution of the elements.

The stars have been classed into four groups, which, as Professor Clarke has remarked, indicate different stages in the process of evolution. The first class, containing white stars like Sirius, show the predominance of hydrogen and a scarcity of the metallic elements. In the second class the metallic elements become more numerous and the hydrogen less distinct; while in the third class hydrogen is difficult to detect.[1] This seems to show a gradual development of the chemical elements as the star cools and grows older. I shall now give a brief account of the views expressed on the subject by some of our leading physicists and chemists.

It will be observed, in reference to the theories we have just considered, that the process of evolution is supposed to take place from the smaller to the larger aggregates of matter. Beginning with an extreme condition of tenuity, by aggregation, the materials become successively larger and more complex. In passing backwards in the process we find the aggregates becoming less and less till they reach the " cosmical dust," or " fire-mist," out of which the primitive nebulæ were supposed to be formed. Receding still farther back, we have the universal

[1] See also on this point Mr. Lockyer's " Bakerian Lecture," *Proc Roy. Soc.* No. 266, p. 21.

atmosphere from which the fire-mist is assumed to have been derived.

This universal atmosphere, though in a state of extreme tenuity, is, as we shall see, supposed by some to be in a more elemental form than anything revealed to us in the laboratory. The suggestion of the dissociation of the chemical elements and its application to stellar physics was, I think, first advanced by Sir Benjamin Brodie in 1866, and more fully in 1867. In the latter year views similar were considered more fully by Dr. T. Sterry Hunt. The question of the dissociation of elements has been ably discussed by Mr. Lockyer in his various writings. It has been suggested by Mr. Lockyer that the coincidence of rays emitted by different chemical elements when subjected to very high temperatures affords evidence of a common element in the composition of the metals producing the coincident rays. Mr. Lockyer states that many trains of thought suggested by solar and stellar physics point to the hypothesis that the *elements themselves, or at all events some of them, are compound bodies.*[1] This view was also put forward by Professor Graham, who says " that it is conceivable that the various kinds of matter now recognised in different elementary substances may possess one and the same element or atomic molecule existing in different conditions of mobility. The essential unity of matter," he adds, " is an hypothesis in harmony with the equal-action of gravity upon all bodies." Similar views have been advocated by M. Dumas, who based

[1] *Proc. Roy. Soc.* vol. xxviii. p. 160.

the suggestion of the composite nature of the elementary atoms on certain relations of atomic weights. The composite nature of the chemical elements has also been maintained by Henri Sainte-Claire Deville, and also by Berthelot, who held that the atoms of the elements are the same, and distinguished only by their modes of motion. Professor Schuster, in a paper read before the British Association in 1880, supports the view of the dissociation of the chemical elements.

That all the purely physical sciences will one day be brought under a few general laws and principles, and the whole of the recognised chemical elements will be resolved into one or two material elements, is a conclusion towards which physical science seems at present slowly tending. There is certainly something fascinating in this view of the unity of nature. There is in this idea more than a purely physical interest attached to it. It has, as I hope to show in a future work, an important bearing on questions relating to the foundations of the true theory of evolution.

The question of the unity of the chemical elements is one, however, yet in a hypothetical condition. Professors Liveing and Dewar, who have given attention to this subject, say : " The supposition that the different elements may be resolved into simpler constituents, or into a single one, has long been a favourite speculation with chemists ; but, however probable this hypothesis may appear *à priori*, it must be acknowledged that the facts derived from the most powerful method of analytical investigation yet devised give it scant support." [1]

Sir Benjamin Brodie on the pre-nebular condition of matter.—There are, considers Sir Benjamin Brodie, very forcible reasons which lead us to suspect that chemical substances are really composed of a primitive system of elemental bodies, analogous in their general nature to our present elements : that some of those bodies which we speak of as elements may be compounds. These ideal elements assumed by him, he says, " though now revealed to us by the numerical properties of chemical equations only as *implicit and dependent existences*, we cannot but surmise may sometimes become, or may in the past have been, *isolated and independent existences* "—as, for instance, in the case of the sun, where the temperature is excessive. " We may," he further adds, " consider that in remote ages the temperature of matter was much higher than it is now, and that these other things [ideal elements] existed then in the state of perfect gases—separate existences—uncombined." [2] He then refers to certain observations of Mr. Huggins and Dr. Miller on the spectra of nebulæ where one of the lines of nitrogen was found alone ; and that this suggested to them that the line might have been produced by one of the elements of nitrogen ; and that nitrogen may therefore be compound. He mentions as a significant fact that a large proportion of the class of elements which he has termed " composite elements " has not been

[1] *Proc. Roy. Soc.* vol. xxxii. p. 230.
[2] *Ideal Chemistry*, p. 56.

found in the sun, they having probably been decomposed by the intense heat.

Dr. T. Sterry Hunt on the pre-nebular condition of matter.—A year after the foregoing views regarding chemical dissociation had been advanced by Sir Benjamin Brodie, Dr. T. Sterry Hunt, in a lecture on " The Chemistry of the Primeval Earth," delivered at the Royal Institution (May 31, 1867), put forward, apparently quite independently, opinions on dissociation similar to those of Brodie. In this lecture he says : " I considered the chemistry of nebulæ, sun, and stars in the combined light of spectroscopic analysis and Deville's researches on dissociation, and concluded with the generalisation that the breaking-up of compounds, or dissociation of elements, by intense heat is a principle of universal application, so that we may suppose that all the elements which make up the sun, or our planet, would, when so intensely heated as to be in the gaseous condition which all matter is capable of assuming, remain uncombined; that is to say, would exist together in the state of chemical elements, whose further dissociation in stellar or nebulous masses may even give us evidence of matter still more elemental than that revealed in the experiments of the laboratory, where we can only conjecture the compound nature of many of the so-called elementary substances." [1] And in his address at the grave of Priestley he referred to the suggestion of Lavoisier that hydrogen, nitrogen, and oxygen, with heat and light, might be regarded as simpler

[1] _American Journal of Science_, vol. xxiii. p. 124.

forms of matter from which all others are derived. This suggestion was considered in connection with the fact that the nebulæ, which we conceive to be condensing into suns and planets, have hitherto shown evidences only of the presence of the first two of these elements, which, as is well known, make up a large part of the gaseous envelope of our planet, in the forms of air and aqueous vapour. With this he connected the hypothesis advanced by Grove, " that our atmosphere and ocean are but portions of the universal medium which, in an attenuated form, fills the interstellary spaces ; [1] and further suggested as a legitimate and plausible speculation that these same nebulæ and their resulting worlds *may be evolved by a process of chemical condensation from this universal atmosphere,* to which they would sustain a relation somewhat

---

[1] " Our atmosphere," says Dr. Hunt," is not terrestrial, but cosmical, being a universal medium diffused throughout all space, but condensed around the various centres of attraction in amount proportional to their mass and temperature, the waters of the ocean themselves belonging to this universal atmosphere." (*Nature,* August 29, 1878, p. 475.) Similar views have been advocated by Mr. Mattieu Williams who says " that the gaseous ocean, in which we are immersed, is but a portion of the infinite atmosphere that fills the whole solidity of space; that links together all the elements of the universe, and diffuses among them their heat and light, and all the other physica and vital forces which heat and light are capable of generating.' (*Fuel of the Sun,* p. 5.) In 1854 Sir William Thomson suggested the idea that the luminiferous ether was probably a continuation of our atmosphere, though I do not think he continues to hold that opinion. The first to advance this idea was, undoubtedly, Newton, who assumed interplanetary space to be universally filled with an ethereal medium " much of the same constitution as air, but far rarer, subtler, and more elastic."

analogous to that of clouds and rain to the aqueous vapour around us."

Professor Oliver Lodge on the pre-nebular condition of matter.—Some have gone still farther back and supposed that the material universe may have arisen out of the luminiferous ether—the hypothetical medium which is assumed to pervade all space. The universal world-stuff scattered through boundless space may in an extreme state of attenuation be, says Professor Winchell, the ethereal medium, and out of this semi-spiritual substance may have germinated the molecules of common matter. "It is certainly possible," he says, "to conceive these cosmical atoms as a rising-out of some transformation of the ethereal medium ; but we know too little of the nature of ether to ground a scientific inference of this kind." [1]

The ethereal origin of matter has been advocated by M. Saigey, Dr. Macvicar, and others. In a lecture by Professor Oliver Lodge, delivered at the London Institution in December 1882, he also advocates the ethereal origin of matter. "As far as we know," to state his views in his own words, "this ether appears to be a perfectly homogeneous, incompressible, continuous body, incapable of being resolved into simple elements or atoms ; it is, in fact, continuous, not molecular. There is no other body of which we can say this, and hence the properties of ether must be somewhat different from those of ordinary matter." . . . "One naturally asks, is there any such clear distinction to be drawn between ether

[1] *World Life*, p. 533.

and matter as we have hitherto tacitly assumed? May they not be different modifications, or even manifestations, of the same thing?" He then adopts Sir William Thomson's theory of vortex atoms, into the details of which I need not here enter. In conclusion, says Professor Lodge, " I have now endeavoured to introduce you to the simplest conception of the material universe which has yet occurred to man— the conception that it is of one universal substance, perfectly homogeneous and continuous, and simple in structure, extending to the farthest limits of space of which we have any knowledge, existing equally everywhere : some portions either at rest or in simple irrotational motion, transmitting the undulations which we call light ; other portions in rotational motion—in vortices, that is—and differentiated permanently from the rest of the medium by reason of this motion.

" These whirling portions constitute what we call matter ; their motion gives them rigidity, and of them our bodies and all other material bodies with which we are acquainted are built up.

" One continuous substance filling all space, which can vibrate as light ; which can be sheared into positive and negative electricity ; which in whirls constitutes matter ; and which transmits by continuity, and not by impact, every action and reaction of which matter is capable. This is the modern view of ether and its functions." [1]

There is this objection to Professor Lodge's theory: it is purely hypothetical. The vortex atoms are not

[1] *Nature*, February 1, 1883, p. 330.

only hypothetical, but the substance out of which these atoms are assumed to be formed is also itself hypothetical. We have no certain evidence that such a medium as is thus supposed exists, or that a medium possessing the qualities attributed to it could exist. In fact, we have here one hypothesis built upon another.

The vortex theory appears to me to be beset by a difficulty of another kind, viz. that of reconciling it with the First Law of Motion. According to that law no body possessing inertia can deviate from the straight line unless forced to do so. A planet will not move round the sun unless it be constantly acted upon by a force deflecting it from the straight path. A grindstone will not rotate on its axis unless its particles are held together by a force preventing them from flying off at a tangent to the curve in which they are moving. Centrifugal force must always be balanced by centripetal force. The difficulty is to understand what force counterbalances the centrifugal force of the rotating material of the vortex-atom. It is not because the centrifugal tendency of the rotating material is controlled by the exterior incompressible fluid, for it offers no resistance whatever to the passage of the atom through it — in short, in so far as the motion of the atom is concerned, this fluid is a perfect void. Now, if this fluid can offer no resistance to the passage of the atom as a whole, how then does it manage to offer such enormous resistance to the materials composing the atom, so as to continually deflect them from the straight path and compel them to move in a curve ? The centrifugal force of these vortex-atoms must be enormous, for on it is assumed to depend the hardness or resistance of matter to pressure. Now the centripetal force which balances this centrifugal force must be equally enormous. If, then, this perfect fluid outside the vortex-atom can exert this enormous force on the revolving material without being itself possessed of vortex-motion, there does not seem to be any necessity for vortex-motion in order to produce resistance. In short, how is the existence of the atom possible under the physical conditions assumed in the theory ? How this may be, like the space of four dimensions, may be expressed in mathematical language, but like it, I fear, it is unthinkable as a physical conception.

Mr. William Crookes on the pre-nebular condition of matter.—In his opening address before the Chemical Section of the British Association in 1886, Mr. William Crookes entered at considerable length into the question of the genesis and evolution of the chemical elements. I shall here give a brief statement of his views as embodied in his important address, and this I shall endeavour to do as nearly as possible in Mr. Crookes's own words.

"We ask," says Mr. Crookes, "whether the chemical elements may not have been evolved from a few antecedent forms of matter—or possibly from only one such—just as it is now held that all the innumerable variations of plants and animals have been developed from fewer and earlier forms of organic life : built up, as Dr. Gladstone remarks, from one another according to some general plan. This building up, or evolution, is above all things not fortuitous: the variation and development which we recognise in the universe run along certain fixed lines which have been preconceived and foreordained. To the careless and hasty eye design and evolution seem antagonistic; the more careful inquirer sees that evolution, steadily proceeding along an ascending scale of excellence, is the strongest argument in favour of a preconceived plan."

Now, as in the organic world, so in the inorganic, it seems natural to view the chemical elements not as primordial, but as the gradual outcome of a process of development, possibly even of a struggle for existence. But this evolution of the elements must have taken place at a period so remote as to be difficult to grasp by the imagination, when our earth, or rather the matter of which it consists, was in a state very different from its present condition. The epoch of elemental development, remarks Mr. Crookes, is decidedly over, and it may be observed that in the opinion of not a few biologists the epoch of organic development is verging upon its close.

Is there then, in the first place, any direct evidence of the transmutation of any supposed "element" of our existing list into another, or of its resolution into anything simpler? To this question Mr. Crookes answers in the negative.

We find ourselves thus driven to indirect evidence —to that which we may glean from the mutual relations of the elementary bodies. First, we may consider the conclusion arrived at by Herschel, and pursued by Clerk-Maxwell, that atoms bear the impress of manufactured articles. " A manufactured article may well be supposed to involve a manufacturer. But it does something more : it implies certainly a raw material, and probably, though not certainly, the existence of by-products, residues, paraleipomena. What or where is here the raw material ? Can we detect any form of matter which bears to the chemical elements a relation like that of a raw material to the finished product, like that, say, of coal-tar to alizarin ? Or can we recognise any elementary bodies which seem like waste or refuse ? Or are all the elements, according to the common view, co-equals ? To these questions no direct answers are forthcoming."

Argument from Prout's Law.— The bearing of the hypothesis of Prout in relation to the evolution of the elements is first considered by Mr. Crookes. If that hypothesis were demonstrated it would show that the accepted elements are not co-equal, but have been formed by a process of expansion or evolution. According to this hypothesis the atomic weights of the elements are multiples by a series of whole numbers of the atomic weight of hydrogen. It is true that accurate determinations of the atomic weights of different elements do not by any means harmonise with the values which Prout's Law requires ; nevertheless the agreement in so many cases is so close that one can scarcely regard the coincidence as accidental.

The atomic weights have been recalculated with extreme care by Professor F. W. Clarke, of Cincin-

nati, and he says that " none of the seeming exceptions
are inexplicable. In short, admitting half-multiples
as legitimate, it is more probable that the few apparent exceptions are due to undetected constant errors
than that the great number of close agreements should
be merely accidental." In reference to this suggestion
of Professor Clarke, Mr. Crookes thinks that it places
the matter upon an entirely new basis. For, suppose
the unit atom to be not hydrogen, but some element
of still lower atomic weight, say *helium*, an element
supposed by many authorities to exist in the sun
and other stellar bodies—an element whose spectrum
consists of a single ray, and whose vapour possesses
no absorbent power, which indicates a remarkable
simplicity of molecular constitution. Granting that
helium exists, all analogy points, says Mr. Crookes, to
its atomic weight being below that of hydrogen ; and
here, then, we have the very element with atomic
weight half that of hydrogen required by Professor
Clarke as the basis of Prout's Law.

Argument from the earth's crust.—The probable
compound nature of the chemical elements, Mr.
Crookes thinks, is better shown by a consideration of
certain peculiarities in their occurrence in the earth's
crust. " We do not," he says, " find them evenly distributed throughout the globe. Nor are they associated in accordance with their specific gravities :
the lighter elements placed on or near the surface,
and the heavier ones following serially deeper and
deeper. Neither can we trace any distinct relation
between local climate and mineral distribution. And

by no means can we say that elements are always or chiefly associated in nature in the order of their so-called chemical affinities : those which have a strong tendency to form with each other definite chemical combinations being found together, whilst those which have little or no such tendency exist apart. We certainly find calcium as carbonate and sulphate, sodium as chloride, silver and lead as sulphides ; but why do we find certain groups of elements, with little affinity for each other, yet existing in juxtaposition or commixture ? "

As instances of such grouping he mentions nickel and cobalt ; the two groups of platinum metals ; and the so-called " rare earths," existing in gadolinite, samarskite, &c. Why, then, are these elements so closely associated ? What agency has brought them together ? It cannot be considered that nickel and cobalt have been deposited in admixture by organic agency ; nor yet the groups of iridium, osmium, and platinum ; ruthenium, rhodium, and palladium.

These features, Mr. Crookes thinks, seem to point to their formation severally from some common material placed in conditions in each case nearly identical.

Argument from the compound radicals.—A strong argument in favour of the compound nature of the elements, Mr. Crookes thinks, is derived from a consideration of their analogy to the compound radicals, or pseudo-elements as they might be called. It may be fairly held that if a body known to be compound is found behaving as an element, this fact lends plausibility to the supposition that the elements are not absolutely simple. From a comparison of the physical properties of inorganic with those of organic compounds, Dr. Carnelley concluded that the elements, as a whole, are analogous to the hydro-carbon radicals. This conclusion, if true, he added, should lead to the further inference that the so-called elements are not truly elementary, but are made up of at least two absolute elements, which he named provisionally A and B.

In Dr. Carnelley's scheme all the chemical elements save hydrogen are supposed to be composed of two simpler elements, $A = 12$ and $B = 2$. Of these he regards A as a tetrad identical with carbon, and B as a monad of negative weight ; perhaps the ethereal fluid of space. His three primary elements are, therefore, carbon, hydrogen, and the ether.

Argument from polymerisation.—The polymeristic theory of the genesis of the chemical elements propounded by Dr. Mills falls next to be considered.

It has been suggested by Dr. E. J. Mills that the pristine matter was once in an intensely heated condition, and that it has reached its present state by a process of free cooling, and that the elements, as we now have them, are the result of successive polymerisations. Chemical substances in cooling naturally increase in density, and we sometimes observe that as the density increases there are critical points corresponding to the formation of new and well-defined substances. The bodies thus formed are known as polymers. From a study of the classification of the elements Mr. Mills is of opinion that the only known polymers of the primitive matter are arsenic, antimony, and perhaps erbium and osmium.

Argument from the Periodic Law.—Lastly a scheme of the origin of the elements, suggested to Mr. Crookes by consideration of Professor Reynolds's method of illustrating the periodic law of Newlands, is discussed.

It was pointed out by Newlands that atomicity and other properties of some of the chemical elements depend on the order in which their atomic weights succeeded one another; and when this law was extended by Professor Mendelejeff to all elements it was apparent that a mathematical relation exists between the elements. This far-reaching law has been fruitful of results. Referring to Professor Reynolds's diagram illustrating the law, Mr. Crookes says : " The more I study the arrangement of this zigzag curve, the more I am convinced that he who grasps the key will be permitted to unlock some of the deepest mysteries of creation. Let us imagine if it is possible to get a glimpse of a few of the secrets here hidden. Let us picture the very beginnings of time, before geological ages, before the earth was thrown off from the central nucleus of molten fluid, before even the sun himself had consolidated from the original *protyle*.[1] Let us still imagine that at this primal stage all was in an ultra-gaseous state, at a temperature inconceivably

---

[1] *Protyle* is the term adopted by Mr. Crookes to designate the original primal matter existing before the evolution of the chemical elements, and out of which they were evolved. Protyle in chemistry is analogous to *protoplasm* in biology, with this difference, however, that protyle is as yet hypothetical, whereas protoplasm is known to be real.

hotter than anything now existing in the visible universe; so high, indeed, that the chemical atoms could not yet have been formed, being still far above their dissociation-point. In so far as *protyle* is capable of radiating or reflecting light, this vast sea of incandescent mist, to an astronomer in a distant star, might have appeared as a nebula, showing in the spectroscope a few isolated lines, forecasts of hydrogen, carbon, and nitrogen spectra.

" But in course of time some process akin to cooling, probably internal, reduces the temperature of the cosmic *protyle* to a point at which the first step in granulation takes place; matter as we know it comes into existence, and atoms are formed. As soon as an atom is formed out of *protyle* it is a store of energy, potential (from its tendency to coalesce with other atoms by gravitation or chemically) and kinetic (from its internal motions). To obtain this energy, the neighbouring *protyle* must be refrigerated by it, and thereby the subsequent formation of other atoms will be accelerated. But with atomic matter the various forms of energy which require matter to render them evident begin to act; and, amongst others, that form of energy which has for one of its factors what we now call *atomic weight*. Let us assume that the elementary *protyle* contains within itself the potentiality of every possible combining proportion or atomic weight. Let it be granted that the whole of our known elements were not at this epoch simultaneously created. The easiest formed element, the one most nearly allied to the *protyle* in simplicity, is

H

first born. Hydrogen—or shall we say helium ?—of all the known elements the one of simplest structure and lowest atomic weight, is the first to come into being. For some time hydrogen would be the only form of matter (as we know it) in existence, and between hydrogen and the next formed element there would be a considerable gap in time, during the latter part of which the element next in order of simplicity would be slowly approaching its birth-point : pending this period we may suppose that the evolutionary process, which soon was to determine the birth of a new element, would also determine its atomic weight, its affinities, and its chemical position."

Professor F. W. Clarke on the pre-nebular condition of matter.—Views on elemental evolution almost similar to those of Mr. Crookes's have been advocated by Professor Clarke. Spectroscopic phenomena, says Professor Clarke, are quite in harmony with the idea that all matter is at bottom one, our supposed atoms being really various aggregations of the same fundamental unit.

" Everybody knows that the nebular hypothesis, as it is to-day, draws its strongest support from spectroscopic facts. There shine the nebulæ in the heavens, and the spectroscope tells us what they really are, namely, vast clouds of incandescent gas, mainly, if not entirely, hydrogen and nitrogen. If we attempt to trace the chain of evolution through which our planet is supposed to have grown, we shall find the sky is full of intermediate forms. The nebulæ themselves appear to be in various stages of development ; the

fixed stars or suns differ widely in chemical constitution and in temperature; our earth is most complex of all. There are no 'missing links' such as
the zoologist longs to discover when he tries to explain the origin of species. First, we have a nebula
containing little more than hydrogen, then a very hot
star with calcium, magnesium, and one or two other
metals added; next comes a cooler sun in which free
hydrogen is missing, but whose chemical complexity
is much increased; at last we reach the true planets
with their multitudes of material forms. Could there
well be a more straightforward story? Could the
unity of creation receive a much more ringing em-
phasis? We see the evolution of planets from
nebulæ still going on, and parallel with it an evolution of higher from lower kinds of matter.

"Just here, perhaps, is the key to the whole subject. If the elements are all in essence one, how
could their many forms originate save by a process
of evolution upward? How could their numerous
relations with each other, and their regular serial ar-
rangements into groups, be better explained? In this,
as in other problems, the hypothesis of evolution is
the simplest, most natural, and best in accordance
with facts." [1]

*Dr. G. Johnstone Stoney on the pre-nebular condi-
tion of matter.*—Further evidence that all the chemical elements were probably evolved from one common
source, is furnished by Dr. G. Johnstone Stoney's

[1] *Popular Science Monthly* for February 1876. See also the
January number for 1873.

"Logarithmic Law of Atomic Weights," a theory
recently advanced in a communication to the Royal
Society.[1] A cardinal feature of this investigation is
that in it atomic weights are represented by volumes,
not by lines. A succession of spheres are taken whose
volumes are proportional to the atomic weights, and
which may be called *the atomic spheres*. When the
radii of these spheres are plotted down on a diagram
as ordinates, and a series of integers as abscissas, the
general form of the logarithmic curve becomes apparent; and close scrutiny has shown that either the
logarithmic curve, or some curve lying very close to
it, expresses the real law of nature.

If, as seems probable, the logarithmic law is the
law of nature, there appear to be three elements
lighter than hydrogen, which Dr. Stoney has termed
infra-fluorine, infra-oxygen, and infra-nitrogen. And
there are, at all events, six missing elements between
hydrogen and lithium.

Dr. Stoney's investigation is based on the fact that
if the atomic weights of the chemical elements be
arranged in order of magnitude, periodic laws come
to light, viz. : those discovered by Newlands, Mendelejeff, and Meyer. From this it follows that there
must be some law connecting the atomic weights with
the successive terms of a numerical series—either
alone or along with other variables.

" This law," says Dr. Stoney, " may be obtained in
one of its graphical forms by plotting down a series
of integers as abscissas, and the successive atomic

[1] *Proc. Roy. Soc.* for April 19, 1888, p. 115.

weights as ordinates. In this way it furnishes a diagram which has somewhat the shape of a hurling-
stick, consisting of a short curved portion succeeded by
a long and nearly straight portion. But as this diagram
cannot be directly identified with any known curve,
it does not suffice for the determination of the law.

" The diagram, however, assumes a form which can
be interpreted when we use the cube roots of the
atomic weights for its ordinates, instead of the atomic
weights themselves. This is equivalent to taking
volumes instead of lines to represent the atomic
weights. When this is done, the ends of the ordinates
are found to lie near a regular and gradual curve,
from which they deviate to the right and left by dis-
placements that are small and appear to follow periodic
laws which have been in part traced. The central
curve is found on examination to be either a logarithmic curve or some curve lying exceedingly close to it.
If the curve be in reality the logarithmic curve, it
furnishes us the law that :

" The cube root of the nth atomic weight $= \kappa \log (n\ q)$
$+$ a small periodic correction ; where κ and q are
constants, the values of which are furnished by the
observations.

" Either this logarithmic law, or a law that lies exceedingly close to it, must be the law of nature."

Referring to this theory, Professor Reynolds says :
" It certainly introduced points of extraordinary importance, though perhaps at present they could not all
quite realise its fullest import. There were several
points of some little difficulty to be grappled with, but

it clearly pointed to the conclusion that we were fast approaching the time when physicists—both chemical and physicists proper—are combining to evolve out of the scientific work lying on the borderland most important and startling facts."

The bearing which Dr. Stoney's conclusions, like those of Mr. Crookes, have on the primitive condition of the material universe is obvious.

Dr. Stoney, like Mr. Crookes, considers that the chemical elements are subject to decay. That they are not only generated but destroyed—that they are subject not only to evolution but dissolution. He believes that the generative process probably takes place only at, or beyond, the confines of the universe, and the destructive process at the centres of overgrown stars, which is the position of lowest potential. Dr. Stoney thinks that this extinction of the chemical elements in the centre of a star is a cause which limits its size and prevents its overgrowth.

THE IMPACT THEORY IN RELATION TO THE FOREGOING THEORIES OF THE PRE-NEBULAR CONDITION OF MATTER.

In all these theories, as has already been observed, the primitive condition of the universe was that of matter in a state of extreme tenuity, while by aggregation the materials became successively larger and larger until they assumed the magnitude of suns and planets. For example, according to the meteoric theory, meteorites are formed out of " cosmical dust," " fire-mist," or condensed vapour, and then suns and planets are formed by aggregation from these meteorites. Facts seem, however, to point to the very reverse as being the true course of events.

Meteorites are undoubtedly the fragments of larger masses. It looks more likely that they are, as has already been stated, fragments of stellar masses which have been shattered to pieces by collision, and that this " cosmical dust," from which the meteorites are alleged to have been formed, are simply the dust arising out of the destruction of the masses. After the two bodies had collided and been shattered to pieces, some of the fragments would undoubtedly be projected with a velocity that would carry them beyond the attractive power of the general mass, and thus they would escape being volatilised. These fragments would continue their wanderings through space as meteorites.

I cannot but think that the number, as well as the importance, of these wanderers has been greatly over-estimated. Mr. Lockyer states that Dr. Schmidt, of Athens, found that the mean hourly number of luminous meteors visible on a clear moonless night by one observer was fourteen. Certainly no such quantity is visible in this country. In Scotland, at least, one may often watch night after night under the most favourable conditions without having the good fortune to see a single meteor.

It is, of course, true that the immediately prior condition of a sun or a planet was that of matter in an extremely attenuated or dissociated state. This is essential to the nebular, as well as to the meteoric hypothesis. But it is not with the immediately prior condition that we are at present concerned, but with the primitive, or pre-nebular, condition. Take, for example, the case of the solar nebula, out of which our sun and planets were formed. Was this nebulous mass formed from matter in a state of extreme tenuity, scattered through space and collected together by gravity ? Or did it result from two solid globes shattered to pieces by collision, which were then converted into the nebulous condition by the heat generated from the collision ? It is no doubt true that the analogies of nature would, at first sight, be apt to lead us to the conclusion that the former theory was the more likely of the two, as the larger is generally made by aggregation from the smaller. But a little consideration will show that, in the present case, the weight of this analogy is more apparent than real. The impact theory does not rest upon a purely hypothetical basis. The cause to which it appeals has a real existence. The point of uncertainty is whether the cause actually produces the effect which is attributed to it. We know from observation that there are stellar masses, some of them probably larger than our sun, moving through space with enormous velocities in all directions.[1] According to the ordinary laws of chance, collision at times would be an inevitable result, and when such an event did take place the destruction of the colliding bodies, and their consequent transformation into a

[1] The dark stellar masses which escape observation may be as numerous as those that are visible.

nebulous mass, would, at least in many cases, be a *necessary* result. In fact, we have, in the case of these vast stellar masses, what we know occurs among the invisible molecules of a gas. So far as mere analogy is concerned, the impact theory is just about as probable as the other.

From what has been stated it would follow that in most cases the stellar masses have been formed out of the destruction of pre-existing masses, like the geological formations out of the destruction of prior formations.

The theories do not account for the motion of the stars.—According to all the foregoing theories aggregation and condensation are produced by gravity. The materials dispersed throughout space are drawn together by their mutual attraction, and aggregated round a centre of gravity. Gravitation, although it imparts motion to the materials, can impart no motion of translation to the mass itself. Gravitation cannot, therefore, be the cause of the motion of translation of the mass. The stars are not supposed to be gravitating towards, or around, a great centre of attraction, for they are found moving in straight lines in all directions, which could not be the case if gravity were the cause of their motion. To what cause is their motion, therefore, to be attributed? A meteorite or other small body might be ejected from any system, by the explosive force of heat or some other cause, with a velocity which might carry it into boundless space ; but such could not be the case in regard to a body of the magnitude of a star. No one for a moment could suppose that 1830 Groombridge, for example, moving at the rate of 200 miles a second, is an eject from any system.

According to the impact theory the whole is plain; for this 200 miles per second is simply a part of the untransformed motion of translation which the materials composing the star had from the beginning. In other words, the matter and the motion were eternal, or, what is more probable, as will afterwards be seen, co-existed from creation—not, however, as molecular motion, but as motion of the mass.

The theories do not account for the amount of heat required.— It has been shown that, although the materials of our solar system had fallen together from an infinite distance, it could not have generated heat sufficient to have formed a gaseous nebula extending to the distance of the planet Neptune. Gravitation alone could not, therefore, have been the source from which the nebula obtained its heat. The solar nebula, however, must originally have extended far beyond the orbit of Neptune.

But supposing it could be demonstrated that the heat thus generated was sufficient to have formed a nebula extending to even twice the distance of Neptune, this would not remove the fatal objection to the gravitation theory of the origin of the solar nebula. For the facts, both of geology and of biology, equally show that the sun has been radiating his heat at the present rate for more than twice the length of time that it could possibly have done had gravitation been the source from which the energy was derived.

This objection is alike fatal to the meteoric theory as it is to all other theories which attribute the origin and source of the heat to gravitation.

Evolution of matter.—Our inquiries into stellar evolution do not, however, begin with the consideration of a gaseous nebula, or with swarms of meteorites. There was a pre-nebular evolution. The researches of Prout, Newlands, Mendelejeff, Meyer, Dumas, Clarke, Lockyer, Crookes, Brodie, Hunt, Graham, Deville, Berthelot, Stoney, Reynolds, Carnelley, Mills, and others, clearly show, I think, that the very matter forming this nebulous mass passed through a long anterior process of evolution. And not only the matter, but the very elements themselves constituting the matter, were evolved out of some prior condition of substance.

I have already given at some length the views which have been advanced by several of our leading physicists and chemists on the evolution of the chemical elements, and on some of the bearings which these views have on stellar evolution. I shall now briefly refer to a point on which I venture to think the theory discussed in this volume seems to cast some additional light.

If the elements were evolved out of a common source, there is, in order to this, one necessary condition, viz. an excessively high temperature ; for the temperature must be above the point of the dissociation of all the chemical elements. "In the primal stage of the universe," says Mr. Crookes, " before matter, as we now find it, was formed from the protyle, all was in an ultra-gaseous state, at a temperature inconceivably hotter than anything now existing in the visible universe ; so high, indeed, that the chemical atoms could not yet have been formed, being still far above their dissociation point."

What, then, produced this excessive temperature in this supposed ultra-gaseous protyle ?   It could not have resulted from condensation by gravity.   In condensation the heat increases as the condensation proceeds, because it is the condensation which produces the heat.   But here the reverse must have been the case, for the ultra-gaseous mass was much hotter than the sun which was afterwards formed out of it. It was, according to Mr. Crookes, when this gaseous mass cooled down, so as to permit of its becoming converted into solid matter, that condensation into a sun could take place.   Besides, was it not the excessive heat which produced the assumed ultra-gaseous condition ?

There is another difficulty besetting the theory that the primitive heat was derived from condensation by gravitation.   Supposing we should assume it possible that the protyle could exist in this ultra-gaseous state without possessing temperature, and that it obtained its heat from condensation by gravity, then the fact of condensation taking place shows that the gas was not in a state of equilibrium.   But the gas could not have remained stationary for a single moment without beginning to condense while in a condition of unstable equilibrium.   We must therefore conclude that the gas must have been in some other condition than the gaseous state prior to condensation.

The impact theory seems to remove all these difficulties. It is just as likely *à priori*, if not more so, that the primitive form of the protyle should have been that of large cold masses moving through space in all directions, with excessive velocities, as that it should have been that of a gaseous mass in a state of unstable equilibrium. If we assume the former condition, then the colliding of these masses would account not only for the ultra-gaseous state, but also for its inconceivably high temperature. Besides, in this case we are not called upon to account for any other antecedent state of the masses before collision, for they may have existed from the beginning of creation in the form of masses in motion through space.

Had space and time permitted, it might have been shown that there are other obscure points on which the theory seems to shed additional light. I shall now, in conclusion, refer to a point wherein the theory differs radically from that of all other theories of stellar evolution. But before doing so I may briefly refer to an objection which has been frequently urged against the theory.

Objection considered.—The objection to which I refer is this, that, had the nebulæ been produced by impact in the way implied in the theory, then we ought to have had some historical record of such an event. I can perceive no force in such an objection. Our historical records, I presume, do not extend much farther back than about 3,000 years, and we have no evidence to conclude that a new nebula makes its appearance in the visible firmament with such fre-

quency; and supposing it did, we have no grounds
for assuming that its production by impact in the
way supposed by the theory would attract general
notice. It is doubtful if the nebula produced
would, in the first instance, be actually visible. I
have shown that the temperature of the nebula could
not have been less than about 300,000,000° C., and
it is very doubtful if the gaseous mass enveloping all
that was solid in the nebula would, at such a temperature, be self-luminous. The probability is that
all the chemical elements composing it would be in
a state of utter dissociation, and converted back into
the original protyle from which they were derived,
again to be slowly reconverted into their former
atomic condition as the temperature fell.

*Can we on scientific grounds trace back the evolu-
tion of the universe to an absolute first condition?*—As
has been repeatedly stated, all inquiries into the
evolutionary history of the stellar universe begin in
the middle of a process. Evolution is a process.
The changes that now occur arose out of preceding
changes, and these preceding changes out of changes
still prior, and so on indefinitely back into the unknown past. This chain of causation—this succession
of change—of consequent and antecedent—could not
in this manner have extended back to infinity, or else
the present stage of the universe's evolution ought to
have been reached infinite ages ago. The evolution
of things must therefore have had a beginning in
time. Professor Winchell, in his final generalisation
to his work, " World Life," has stated this matter so

clearly and forcibly that I cannot do better than here quote his words on the subject.

"We have not," says Professor Winchell, " the slightest scientific grounds for assuming that matter existed in a certain condition from all eternity, and only began undergoing its changes a few millions or billions of years ago. The essential activity of the powers ascribed to it forbids the thought. For all that we know—and, indeed, as the *conclusion* from all that we know—primal matter began its progressive changes on the morning of its existence. As, therefore, the series of changes is demonstrably finite, the life-time of matter itself is necessarily finite. There is no real refuge from this conclusion; for, if we suppose the beginning of the present cycle to have been only a restitution of an older order effected by the operations of natural causes, and suppose—what science is unable to comprehend—that older order to be a similar re-inauguration, and so on indefinitely through the past, we only postpone the predication of an absolute begin-ning, since, by all the admissions of modern scientific philosophy, it is a necessity of nature to run down."

These are consequences which necessarily follow from every theory of stellar evolution which has hitherto been advanced. The impact theory, how-ever, completely removes the difficulty, for according to it the evolutionary process can, on purely scientific grounds, be traced back to an absolute beginning in time. If huge solid masses moving through space were the original condition of the universe, then, in so far as either philosophy or science can demonstrate to the contrary, it might have been in this condition from all eternity. We are therefore not called upon to account for this primitive condition of things. Now it is evident, unless a collision should take place, the universe would remain in this condition for ever: without a collision there could be no change, no work performed, and absolutely no loss or gain of energy, and therefore no process of evolution. The first collision would be the absolute commencement of evolution—the beginning of the process of the development of the universe. Evolution would, in this case, have its absolute beginning in time, and consequently was not eternal. If, on the other hand, we assume, what is far more in harmony with physics, metaphysics, and common sense, that the universe was created in time, we are still led to the same result as to an absolute commencement of evolution. In both cases we reach a point beyond which there can be no legitimate inquiry ; no further question which the scientists can reasonably ask.

We have no grounds to conclude that there is anything eternal, except God, Time, and Space. But if time and space be subjective, as Kant supposes, and not modes pertaining to the existence of things in themselves, then God alone was uncreated, and *of* Him and *to* Him are all things.

INDEX

I

PRINTED BY EDWARD STANFORD
26 AND 27 COCKSPUR STREET, CHARING CROSS, LONDON, S.W.